The Science Of God Volume 3

R Lindemann

Aleph Publications
Wisconsin, USA

The Science Of God Volume 3
Day Five and Day Six - The Creatures - Revolution or Evolution
Copyright 2016 - R Lindemann ©
All Rights Reserved. Published 2023

Aleph Publications
Manitowoc WI

All rights reserved. No part of this publication may be stored in a retrieval system, reproduced, or transmitted in any form, electronic, mechanical, photocopying, recording or other, without first obtaining the written permission of the copyright owners and the publisher.

Paperback Edition
ISBN13: 978-1-956814-28-6

33 32 31 30 29 28 27 26 25 24   2 3 4 5 6

Disclaimer

All information, views, thoughts, and opinions expressed herein are those of the author(s) and are being presented only for your consideration and should not be interpreted as advice to take any action. Any action you take with regard to implementing or not implementing the information, views, thoughts, and opinions contained within this published work is your own responsibility. Under no circumstances are distributor(s) and/or publisher(s) and/or author(s) of this work liable for any of your actions.

Anyone, especially those who have been victim of misdirected explanation and understanding, may be best served seeking wise counsel before deciding to implement any information, views, thoughts, opinions, or anything else that is offered for your consideration in this work. All information, views, thoughts, and opinions in this work are not advice, directive, recommendation, counsel, or any other indication for anyone to take any action. All information, views, thoughts, and opinions offered herein are offered only as suggestions for your personal consideration, which is done of your own free will. Your life is your own responsibility; use it wisely.

Any use of trade names or mention of commercial sources is for informational purposes only and does not imply endorsement or affiliation.

Please note that most of the items in quotes in this book are from various versions of the Bible and may have been paraphrased.

Dedication

To all those who seek logical explanation about the origins of animal life and wonder about the viability of both the Biblical account and the evolution theories, and to all those who have made great discoveries regarding animal origins, this book is dedicated to you! Without people wondering, and without the previous thinkers, we would not know or understand that which we understand today. Whether their ideas were right or wrong, we all have benefited from their work to more clearly know, today, what is and what is not even though we have a long way yet to go.

Contents

Chapter 1
The Beginning of DNA ... 1
- Create ... 2
- The First Cell ... 3
- What Does the Bible Actually Say About Creation? ... 4

Chapter 2
Let the Waters Bring Forth ... 9
- The Chicken or the Egg? ... 11
- Similarities In Embryos ... 16
- Brachiopods ... 18
- Archaeopteryx the Dinosaur ... 19

Chapter 3
Let the Earth Bring Forth ... 21
- Elevate Your Thinking ... 22
- Animals On High Ground ... 23
- Defining Animals ... 24

Chapter 4
Making a Way Using Scientific Extrapolation ... 27
- Extrapolating Finches to Crocodiles ... 28
- Evolution Fails the Scientific Method ... 28
- Is It Really Demonstrable? ... 29
- The Paper Trail a Mile High ... 30
- Evolutionary Laws? ... 33

Chapter 5
At First Breath ... 35
- Interpretation Manipulation ... 36
- Analysis of Terms ... 40
- Define Terms ... 41
- Does a Science Dictionary Exist? ... 44

Chapter 6
Two of a Kind Makes a Full House ... 47
- Kinds ... 47
- Why Versus How ... 50
- Common Ancestors In a Single Cell ... 51
- Two of Every Kind ... 52

Chapter 7
Making the Right Separations ... 55
- "cus" Words of Phylogeny ... 56
- Systematic Classification of Life ... 57
- Defining Phylogeny ... 59
- Challenge Phylogeny ... 59
- Common Ancestor ... 61

Chapter 8
The Bodies of Evidence ... 63
- Biblesaurus ... 63
- Scientific History ... 68
- The Christian and How It All Came to Be ... 70
- Dinosaur Pictures and Other Figures ... 71

Chapter 9
The Big and Small 73
    Scaling Things Down 74
    Pollywogs 74
    Location, Location, Location 75
    Bird, Brains, and Bones 76

Chapter 10
Evolution of Species 77
    Families and Kingdoms Don't Exist?..77
    Finding New Evidence 78
    Definition of Species 79
    Stop Crossing Species 80
    Blatantly Dishonest 81

Chapter 11
Receiving the DNA 83
    Walking Fish...83
    Genetic Language 85
    Lost Information 86
    What We Do and Don't Know About DNA 86

Chapter 12
What Guides the Changes 89
    Moving Theories 89
    Evolved from a Rock? 91
    Logical Contradictions 91
    Evolving Parts 92

Chapter 13
Life Forms Are Fluid 93
    The Oceans Brought Forth 93
    Descent from Primitive Life 95
    Unused Appendages..95
    Bears and Dogs Connected 96

Chapter 14
Life is Intended to Be Robust 99
    White Bear 100
    Adapting to the Circumstances 101

Chapter 15
Like with Like 103
    What is a Kind? 104
    Why Are Zebras Stripped? 106
    Beauty In Nature 108
    Creature Groups 109

Chapter 16
The Elements of Life 113
    Goldilocks Planet. 114
    Relativism Abounds 115
    What is Truth? 115
    Basic Building Blocks 116

Chapter 17
Do the Fossils Tell the Whole Story? 119

Tree of Life	122
Mono Phylogeny	123
Don't Be Fooled	124
Reclassifying Skeletal Finds	125

Chapter 18
What Do You Really Want to Know 127
Did Aliens Bring Life Fully Formed to Earth?	128
Murchison Meteorite	130
No One Doubts	131
The Laws of Science	132

Chapter 19
Everything Has Been Saved for You 135
Selective Listening versus Selective Believing	136
Believing Wrongly	137
A Mountain of Evidence	138

Chapter 20
Is There Any Meaning to It All? 141
What is "proof"	142
Order is Random	144
Superficial Hypothesis	146
Darwinism	147

Chapter 21
The Darwin Delusion 149
Animal Intellect	150
Evolutionism is Brainwashing	151
Darwin's Evolution Superstitions	152
Darwinism is Foul and Dishonest	153
Will Unbelievers Suffer Hell?	154
Darwin is Their God	155
Floundering Evolution	155
Evolution is a Belief System	156
Evolutionism Opinion versus Facts	157
Pity Evolutionists	158
Darwin's Task List	159

Chapter 22
Choosing Partners 161
No Really, Trust Me	162
The Rotting Tree of Life	163
Cladograms	164
Undeniable Connections	165

Chapter 23
Why Didn't It Happen Some Other Way? 167
What Are Eukaryotes	168
Law or Principle of Monophyly?	168
The True Definition of "kind"	170
Hung Up On Scientific Terms	171

Chapter 24
Eternal Consistency 173
Impotent or Omnipotent	175
Science is Retreating from Science	176

The Shallow Science of Evolutionism	177
We Come to a Point of Belief	178
Points to Consider Rather than to Ignore	179
Final Thoughts to Consider	182
Is There An Answer?	182
How Did Creatures Get Here?	186
The Bringing Forth	191
Was it Evolution or was it Creation?	193
What Came First, Archaeopteryx or the Egg?	196
Eggsactly	198
Shared Attributes	200
The Spark of Life	202

Acknowledgements

Thank you to all who have studied the subject of animal origins, both past and present thinkers. Your collective information is a treasure-trove of insight. A thought proven or disproven is of great value to everyone alive today who questions the origins of life.

And special thanks to all who have assisted in discussion and critical editing of this book. The most important factor in being sure about our thoughts is allowing them to be challenged. Doing so forces us to reevaluate our position to make certain our logic is solid, and if not, then adjust our thinking as needed.

Additional thanks to all of those who are rarely acknowledged for their contribution to all authors. For without the inventors and workers making the tires for the trucks that carry the wood from the forest to the people who make pulp for the paper for books and also those that farm the food that those people eat in order to sustain life so that they are able to do their jobs, no author could present their work to the world. Thank you all! Even electronically, the same principle applies, without the technicians that maintain computer infrastructures and without the innovators that designed the components for the entire network, no print or electronic books would be able to be distributed in our modern era. Volumes of books could be written just to thank everyone in this miraculous chain of events! From the people mining the copper for the wire wrapped on the armature of the alternator of the car that drives the worker to his lumber-hauling truck to the people who are drilling for oil and mining coal and maintaining hydroelectric dams that provide the power for the checkout girl to drive to work to accept payment for the work of any author, thanks to you all! You are all invaluable to the world, and you are needed and loved!

Introduction

Whether you believe in evolution or believe in the Bible, you likely have wondered where the animals truly came from and if the explanation you have chosen to believe is possible. There are valid points to consider on both sides of the animal origins topic. Many people believe that the Bible appears to imply that the animals arrived suddenly at the hand of God with no evolution whatsoever, where the evolution model promoted by Darwin removes any thought of assisted creation and relies solely upon the survival-of-the-fittest model. So which is it? Is the Bible accurate? Or is evolution real?

The deeper you look into any of the creation topics the more mixed messages you will encounter. What frequently occurs is that people who believe in Biblical creation will study science, typically in high school and college, and then will attempt to reconcile the Bible with evolution. Some people actually go as far as claiming the Bible's "Adam and Eve" were the first primates that displayed enough intellect to be able to make an account of events and then pass the stories of those events on down from generation to generation. But that is for a later volume, and when it comes to animals there is no need to account for any animal having an intellect anywhere near our human level of intellect. Animals seem to have been animals throughout their entire history and we have no indication otherwise.

Since this book, *The Science Of God Volume 3 - Day Five and Day Six - The Creatures - Revolution or Evolution*, is specifically about animals and other creatures, we need not really be discussing human-like intellect, or humans in general for that matter. What we are specifically working to understand in this book, *Volume 3*, is how might animals have come to be and whether or not a "God" or a Creator could have had any influence on their arrival and form.

To get to the bare bones of this topic, we first have to try to grasp exactly what the Bible says regarding the arrival of animal life, which takes a bit more contemplation than appears on the surface. As we take a deep dive into this subject, we will still be keeping things simple and steer clear of lots of scientific jargon. The goal here is to understand what the Bible actually says and if any of what it actually says is remotely compatible with any evolution or science whatsoever. We also want to consider what science actually says about evolution.

The two general perspectives that society is typically presented with are two diametrically opposed ideas. One is the Biblical argument that God said "Let there be", and Poof! Animals of all sorts! Then on the other end of this tug-of-war rope is the evolutionary side that says that the Earth is three-and-a-half-billion to four-and-a-half-billion years old, at last count, and that animals began as a primitive single-celled organism that morphed into increasingly more complex organisms with each subsequent generation with each new generation continuing to do so even unto today, and all without any guidance whatsoever from any intelligent creative power, which is to say a "God".

So what's the real story of the origin of animals? Is either of the two ends of this creation tug-of-war rope legitimate? Is either accurate as currently conveyed by modern science? If you dive in deep you can find many books written on both sides of the debate "proving" that one or the other is true and accurate. However, I will give a bit more credit to the consistency of the evolutionists in this because they tend to say it is absent of "God" or intelligence of any sort, where the Biblical crowd tends to have a variety of theories that go from entirely God-induced within twenty-four hours, all the way to long-age evolution at the hand of God, as well as just about any imaginable theory in-between.

Few people are open enough to begin reading a book such as this without having their opinion about the book being affected beforehand due to their own preconceptions that are derived from their upbringing, from society, from their education, and from their occupation. But, rather than insisting that this way or that way is the only way, reserve your analysis to the end of the book and then consider what you have read and see then how it fits with, or compares to, the way you previously had the events lined up in your own mind. For most people, much of the information presented in this book will be a new perspective that they likely never considered or ever even heard.

So let's get to it! Just how did all of the creatures come to be? Is it possible that a "God" or Creator had a hand in it? Or does science already have it all accurately figured out? You decide.

Chapter 1

The Beginning of DNA

Don't let this brief talk of "DNA" make you think this is what you are about to read about. Yes, there is mention of DNA in this book, but there are no detailed complexities beyond this basic initial information found here in the first few paragraphs of this chapter.

Is DNA what we think it is? Is DNA really the instruction set to life? And does DNA morph? Or was it made firm and constant on Genesis' days five and six with little or no deviation since then? To sum up the big question, did God do it?

Deoxyribo**N**ucleic **A**cid (or DNA), as far as we can tell is a long and complex series of four simple parts or bases (cytosine "C", guanine as "G", adenine "A" or thymine as "T"). Blah blah blah, right?

With a relative amount of certainty, as a culture, we have scientifically compelling evidence that messing around with this chain of DNA comprised of C, G, A, and T bases does indeed affect the outcome of an organism that will develop from this perceived instruction set. Much like a computer program using sequences of code, these four bases appear to determine things like hair color, eye color, gender, etc. But in the case of animals we are not as concerned about those attributes that are commonly associated with humans. Instead we are questioning if the DNA base-four code (C-G-A-T) can create a program that is so complexly arranged so as to cause the development of the varying animal forms by only slightly deviating from generation to generation.

Create

To begin this discussion, things first need to be organized in to various categories of thought, with the first category being "creation" done at the hand of a "God", that is to say the so-called "intelligent design" argument. In the intelligent design category we must remove the idea of a "God" and be much more specific and state this so-called "God" as a "Creator", which is a discerning spirit being. Additionally, it is very important if we are going to attempt a rational discussion, that we toss aside the old man in a throne of clouds that many people have in their minds regarding "God". This alleged God/Creator created EVERYTHING according to the Bible, and thus the God/source of creation could not have had any visible or tangible form whatsoever during the creation events.

The next category we must recognize is that of god-less evolution. "God-less" is not meant as any sort of demeaning or derogatory comment, but rather is used to clearly illustrate that evolution as proposed by pop-science states that there was zero intelligent input in the formation of life, and it also asserts that life is formed of itself through physical and chemical interaction and reaction.

The First Cell

Now that we have categorized the seeds of creation, we have to try to determine how the first cell might have come to be. When using the hocus-pocus, Poof!, God-did-it! approach, we simply ignore everything or anything that we don't like and blindly credit it all to God. It is a "safe" yet blind approach. However, this approach is contrary to the God of the Bible. We are supposed to try to know this God/Creator. Thus taking a blind and ignorant approach in this way does a disservice to both God and to the Bible.

Jumping to the god-less evolutionary model, we are also faced with a bit of a blind faith issue. We have to imagine that through pure random chance that some sort of electrical charge hit some raw chemicals or gases in just the right way so as to create the first amino acids that then in turn allowed the DNA process to begin.

Then, of course, we have the combination category that is basically crediting God for the DNA regardless of how it might have occurred.

But in all of this we are still somewhat faced with the "what came first, the chicken or the egg" anomaly. But since we are speaking of single cells we are not actually considering a chicken and an egg, but rather that the DNA is what determines the cell. So what came first, the cell or the DNA embodied in the cell? Without the DNA the cell had no instruction from which to derive its form. The DNA and the cell appear to need each other to thrive, or for that matter to exist.

Now on the Biblical side where we have the blind faithed belief that the Creator just did it and it's all settled, there are many people who will accept that, but they still want to know *how* the Creator did it. I do not believe that the *how* question is off limits. If there truly is a God/Creator then that Creator wants us to discover these things.

Regarding the scientific perspective, there are scientists working in their labs trying to create life, but have thus far failed to do so from scratch even though they have countless trillions of cells to copy from and the materials with which to do so. However, they have demonstrated some compelling experiments, such as making what is essentially an empty cell able to move about in fluid, but in most cases these are nothing more than tiny self-propelled balloons, some of which are able to split apart or divide. But this is little more than what we see with children playing with their container of bubbles at a birthday party.

While such scientists should be applauded for their efforts and have in fact used their intelligence to create a truly very incomplete cell, such researchers are nowhere near actually creating "life" in a lab, even with perfect examples to copy from all around them. In the lab we can easily clone things, but those cloning attempts, while successful, still use the entirety of the original cells' functionality and DNA instruction set.

With all of that said we are still faced with coming up with a sound explanation of how that first cell came to be, and in general it does not matter if it was a random chance of electrical charge intermingling with chemicals, or if the Creator did it, because we still want to know *how* it was done. What were the mechanisms that allowed cells to come to be?

Without getting philosophical about it, we all intuitively understand at some point, whether intelligently designed or random chance, that there had to be a first cell, and a part of our quest is to try to figure out *how*. There obviously was a first cell or cells somehow somewhere, with or without God.

What Does the Bible Actually Say About Creation?

If you have read any of the other several volumes of *The Science of God* you should have a pretty good idea of the many ways in which the words of the Bible are twisted by people to achieve the agenda that they want to promote. In fact, it is often those who claim to *not* believe in the Bible or the God of the Bible who tend to profess their knowledge of what the Bible does or does not say regarding creation. Now, this is not a problem in itself since some of these people have studied the Bible in-depth and then arrived at their current god-less view of life and science. But this is where things get a bit sketchy regarding interpretation of the Bible. The book *Understanding The Bible – The Bible How-To Manual* AND *The Things We Don't See* goes into important detail about the Bible and interpretation and the various Bible versions.

There is a great amount of misunderstanding regarding the Bible, and most specifically the book of Genesis, and even more specifically Genesis Chapter One. As explained in *Understanding The Bible*, with the advent of high speed and economical printing methods it became very achievable for someone with ambition to reword the Bible to what they believed to be a more clear phrasing of the Bible's meaning. This is not necessarily a bad thing, but it can be. When looking at the overall message of salvation of the Bible, Genesis One has little effect unless you are interested in the creation subject. What occurred in the seventeen and eighteen hundreds, and even after, is that people could now afford to rewrite the Bible and have it distributed cost effectively to the far reaches of the world which could all be done relatively quickly due to mechanized automated printing.

The problem that we encounter with these newer versions/interpretations of the Bible is that they have altered critical words and phrases, thus forever affecting people's ability to properly interpret the Genesis One creation text. You might think "who cares, just get on with it already", but this is perhaps the single most important oversight that both those who believe the Bible and those who believe science indulge themselves in. In *Volume 1* of *The Science Of God* the ramifications of these more recent translations is explained in greater detail and how it affects our interpretation and how that affected interpretation has affected our perception of the Genesis One events.

When it comes to the animals and other creatures in the Bible, you have to realize that there is very little said about their creation. You also must clearly understand that the arrival of creatures was over two of the so-called "days" detailed in the Genesis creation account—a very important point to keep in mind!

The order of events in the Genesis One creation account is critical to have clearly arranged in your mind in order to be able to rationally discuss the creation text. However, ever since the eighteenth century, and later, the rewrites of the Bible were done, which subsequently caused an improper cementing of many people's view of the text, further causing an unspoken illogical mental rearranging of creation events. It is one of those things that we simply don't see until we see it, and even if someone tells us directly, it will still often be missed by many people due to the fact that the incorrect thinking is too deeply embedded in us to a point where we simply are incapable of seeing past it. That is why in the introduction of this Volume of *The Science Of God* it is asked that you suspend judgment until *after* reading this entire book, and then contemplate it and try to derive any conclusions about the creation account at that point in time.

What the Bible says about creation is really all about the order of events. Unless we can grasp that simple fact, we will be unable to grasp what Genesis is trying to convey to us regarding creation, and in the case of this book we are specifically mostly discussing days five and six regarding the creation of the animals and other creatures. In discussing this topic with most people that I have had the privilege to speak with, it appears that very few people clearly understand the order of events, which is mostly because they don't feel it is important, or they cannot see past the indoctrination that they have been subjected to in life.

The Bible does explain a limited amount of information about the animals and other creatures, but it is really quite vague with one very explicit point made in the text: It says on day five "let the waters bring forth" and on day six "let the earth bring forth". Now these two very explicit statements are really quite important in this entire creation versus evolution discussion. In the description of both days five and six, a couple of phrases later the Genesis text credits God for creating the creatures. This helps to introduce us to *how* God did it while still crediting God for doing the actual creating. Throughout this book we will be splitting some very critical creature hairs regarding the order of creation events and what Genesis actually says.

It is important to note that when creation discussions are occurring you should never use any of the Bible versions that have perverted, altered, or changed the critical words in the Genesis creation account. The Douay Rheims Bible and the King James Bible are both generally very good at having kept true to the original phrasing intent and words used. That is to say respectively, the Catholic and Protestant Bible versions. But let it be known that it is the largely the protestant wing of the Christian religions that have rewritten the Bible and altered the seemingly minor, yet critically important, text of Genesis One in the past couple centuries.

To sidetrack a bit, while it seems frustrating that we have to choose a Bible version to study from, all of these versions are a gift if we use them properly. When doing Bible study, especially when reading the creation text, the words, phrases, and order of events are absolutely critical. Using several key and authoritative versions side by side offers you substantially more certainty that the version(s) you are using are closely phrased and thus more likely accurate. Additionally, it is also wise in our high-tech world to download digital copies/pictures of the original prints of the Douay Rheims Bible and of the King James Bible, and if you are really a stickler for details, consider getting copies of the Guttenberg version of the Douay Rheims and also a copy of the clarified Latin Jerome Vulgate, and to dive even deeper look for the Greek Septuagint as well any other ancient versions of the text that you can find.

While little of that is necessary, you will likely find it quite handy when you are questioning a word or phrase. Regardless, it is best to use a copy of the Bible that was published pre-seventeen hundred, or an accurate derivative of one from that era. You can learn more about the various Bible versions and the details of them in the book *Understanding The Bible – The Bible how-To Manual* AND *The Things We Don't See*.

The words used and the order of events matter, so choose your study Bibles with care, and clear your mind of all indoctrination of both religion and science. Make your mind a clean ready-and-willing, open-minded slate. Reanalyze the information you have collected over the years as you read on!

Chapter 2

Let the Waters Bring Forth

As we get deeper into the evolution/creation debate we must take great care in analyzing the order of events and the specific words used. In its most basic form, what we first need to grasp is that the creatures are broken into two specific groups. The first group was brought forth on "Day Five" in the Genesis text and the second group was brought forth on "Day Six" in the Genesis text. At first glance this might seem meaningless, but there is a very important distinction in what was specifically brought forth on both days, and for this chapter we are generally only concerned with day five.

Douay English Genesis 1 text reads:

"[20] God also said: Let the waters bring forth the creeping creature having life, and the fowl that may fly over the earth under the firmament of heaven.

[21] And God created the great whales, and every living and moving creature, which the waters brought forth, according to their kinds, and every winged fowl according to its kind. And God saw that it was good. [22] And he blessed them, saying: Increase and multiply, and fill the waters of the sea: and let the

birds be multiplied upon the earth. [23] And the evening and morning were the fifth day. [24]"

Consider the broad nature of these phrases:
- Let the waters bring forth the creeping creature having life
- and the fowl that may fly over the earth under the firmament of heaven
- And God created the great whales, and every living and moving creature
- which the waters brought forth
- according to their kinds
- and every winged fowl according to its kind

"Let the waters bring forth" is a scientifically interesting choice of words. This is especially so since the claim of evolution is that lightning theoretically struck some sort of slurry of chemicals that set off that long-term chain-reaction leading to life as we know it today. Yet we are still stuck with the question of twenty-four-hour days versus long-age evolution. Biblically it says "God Said Let the waters... ... and evening and morning were the fifth day." Now we can argue that both the fifth and six days were twenty-four-hour Earth days, but when it comes to the four previous days, namely days one through four, there could be no counting of days as we know them today because there was no Sun shining at that time according to the order of events. The long-days theory tends to ruffle the feathers of the six-twenty-four-hour-day creationists. Technically, in the order of events, the Sun could not have shone until some point during the fourth day. Indicating that no counting of Earth days as we know them today could have existed prior to that moment.

You might wonder about the need for discussing the time frames involved here regarding the fifth day, but it is critically important if we are to ever ascertain what really happened. For instance, if day five was not one of our twenty-four-hour days, then how long might it have been? While I personally don't like to jump around in the Bible too much and I like to keep the study of Biblical text in close proximity to the physical text area and the specific subject being studied, but from a God-perspective, the Bible does state that a day to God is similar to a thousand years for man. From this we should probably not specifically derive that there is an exact equality of one day to one thousand years from God to man respectively, but rather that a thousand years is nothing to God and that a thousand of our years goes by for God in what is like a day to us, meaning that it goes by fast! If you're over forty or fifty years old and you think back on life, you likely have had the realization that the older you get the faster the days seem to go by. Imagine now that if this Creator does truly exist and is "the ancient of days", possibly billions or even trillions or more of our Earth years old, then in noticing our own experience of days flying by as we age, how much more if you were billions or more years old–a day would seem as nothing to you.

What we are trying to establish here is the allotted amount of time the "waters" had to bring forth its respective creatures. Biblically speaking, we know that it would have been a minimum of one of our twenty-four-hour Earth days, but that's really a bit of a stretch when considering that things must grow and mature. We have experienced *nothing* on this Earth to suggest any deviation from the current standard growth model. Which brings us to the next point to try to make sense of–What came first?

The Chicken or the Egg?

So we have vaguely established that the days described in Genesis were likely not specifically the equivalent of our twenty-four-hour Earth days. This is especially true of the first four days of creation.

Regardless of how much time there was for the water to bring forth, we have to first work to logic our way through the creature creation process. What came first, the chicken or the egg? If we cannot come to a consensus on this particular question we will be eternally stuck in a quagmire of creation slurry.

When six-twenty-four-hour-day creationists think this through they are stuck with having to conclude one of a few different possibilities. The first is that God instantly made fully grown chickens capable of breeding, or second, God made eggs fully formed and placed them in the right conditions required to hatch them. Or lastly, that the chickens were formed in some sort of embryonic state and grew quickly in a twenty-four-hour period, both male and female.

Of those three, the first two, being poof! a fully formed chicken, or poof! a fully formed egg, both stretch logic and science beyond their limits.

Let's explore the scientific explanation of this chick/egg issue. Using basic human logic we would initially conclude that the chicken came first because eggs come from chickens. Thus, the chicken had to come first, but chickens hatch from eggs that come from chickens–Enter Evolution: In the evolution model, the chickens began in a primordial soup and may have come out as some sort of worm or amoeba, and slowly over tens or hundreds of thousands or millions or even billions of years they developed into the form that we see them as today. This makes some sense, until we begin to look into the formation of chicken DNA.

We have heard vehement debate on both sides of the DNA argument stating that no information can be added to DNA and that it only loses information. However, when this is stated by evolutionists it is typically after a well-formed creature is already formed, such as certain fossil types over time appearing to have lost a given attribute such as fins for a given creature type. Creationists on the other hand use this argument from inception onward and say that everything was created perfect and only decays over time and through mutation can only lose information.

Regardless, it is really a foolish argument either way because DNA is quite obviously very durable and elastic in that it is capable of changing rapidly—very rapidly! In fact, we humans will take plants or animals and will crossbreed animals or crossbreed plants to create new variations of those animals, and we do the same with plants. Now, one could claim that we are removing information from the DNA when we do this, and that may in fact be true, but we are also altering the very same DNA and infusing into it *new* attributes that it formerly did not contain as is made visibly evident through the offspring of the particular crossbreeding attempt. Because there are so many varying opinions in this debate, on both sides, it needs to be clearly stated and logically understood that DNA must, both, be able to be added to and taken from in order for any sort of evolution or deviation of "kinds" to occur.

This simple crossbreeding example is not mere conjecture; it is absolutely and undeniably proven with each interbreeding attempt that is done. So the question of DNA being able to have information added is not really a question. The real question about DNA is not *if* it can be altered and added to or taken from, but rather what situation or condition is able to affect the DNA in a way where it could alter the offspring built by that DNA?

Briefly getting back to the chicken versus the egg issue, it is likely that it is in essence, both. Not that there was a fully formed egg or a fully formed chicken, but rather that a chicken likely had formed from the cellular state and grew to a fully formed chicken. But let's say for sake of discussion that a chicken did come from some sort of primitive embryo that ultimately formed a chicken from the get go, thus not having had to evolve. Then did one of each gender form, that is to say, a male and a female? This brings to question the idea of pure god-less evolution.

We see cells miraculously multiply by dividing etc... but how or when did the male/female model begin? Did the deviation of the species always breed in a male/female manner? To properly consider this we must back up the evolution train all the way back to the beginning, way back to the first spark that is claimed to have hit the gases causing the first amino acids to form. With these acids it is believed that DNA was able to begin and somehow morph itself into life as we know it today. Is this logical?

Incrementalism reigns supreme in evolution. When you take each detail about life and explain it away through natural selection it seems to be a very plausible explanation. But immediately you then have to explain each incremental detail, and there are too many to count. On the evolution side this is where things tend to get evermore glossed over because with each subsequent incrementalization there are more parts to explain in each subsequent increment. Some proponents of evolution attempt to explain some of these details, but there are too many details to explain to a point where, other than the broad blind-faithed explanation of general evolution, there is no real way to consider all of it in detail. This is especially obvious when you consider that most biologists study only a small area of biology and must gloss over the remainder so that they can do the job they were hired to do.

It appears whether evolution or creation, at some point it requires some level of blind faith in the given belief, though either side of the debate will reject that notion. You can quickly prove this by simply asking "how?" with each explanation offered. If you are making a claim and insisting that it is "fact" then you had best be able to back up that claim to the very end of the extensive sequence of "how?" questions. And if you are unable to offer *factual* answers then you are practicing blind faith beliefs in your science religion.

If something is being promoted as a mere point of contemplation or a basic theory, then there is no need for it to be held to the same standards. But both the six-twenty-four-hour-day creation model and the evolution model are often promoted as "fact" or "absolute" or "absolute fact" or worse "undeniable" and yet much of what they profess as "true" or "fact" is not provable by anything or any evidence that they offer. This would not be a problem with evolution, but in the twenty-first century Darwinian evolution is mistakenly presented as "inarguable fact", which if you take a few moments to logic through, you will find that very critical parts of the theory are patently false or nothing more than mere overzealous speculation.

And the likelihood of a Creator arbitrarily creating ready to go adult creatures on a poof-like manner is low, very low, and the assertion that the days referred to in Genesis One are twenty-four-hour Earth days is highly unlikely, which is expounded on in *The Science Of God Volume 1 – The First Four Days* and *The Science Of God Volume 2 – Day Three - Gravity, Land, Seas, and Evolution of Plants*. Each Volume of *The Science Of God* is grouped into major subject matter, such as

1. Astrophysics–(Volume 1)
2. Arrival of Organic Life–(Volume 2)
3. Arrival of Creatures–(Volume 3)
4. The arrival of Mankind–(Volume 4)
5. The Flood of Noah's Time–(Volume 5)

The sequence of events matters a great deal through all of the stages of Biblical creation. But when we mess around with the text and compromise it, we end up with some very inaccurate perceptions of what Genesis is *actually* trying to convey to us.

"Day five" likely consisted of many of our earth years, which would allow for life to develop in a scientifically acceptable manner. So the chicken likely grew from an embryonic state to full adult in a typical maturation period of current length Earth days and was likely accompanied by the opposite gender chicken/rooster, and likely ready to lay eggs at the typical point of maturity. But this begs the question, did it evolve?

This question ultimately depends on your take on the chicken versus egg question. If what we call a "chicken" today is a variation of another type of bird, then the hard-shell egg, specifically a *chicken* egg, likely came first. Thus, the question can be further whittled down to what came first a bird or an egg? Or in the case of dinosaurs, was it the "Archaeopteryx" or the egg that came first?

Let's examine the basics of a typical egg for a moment. What is an egg? A typical bird egg has a breathable shell made of mostly calcium carbonate and in it is an inner skin, the albumen, and then within that is the yolk. There are no instructions in the egg or mechanical devices that are building the chicken/bird. We can crack open a fresh egg and we will only see liquid, but if that egg was fertilized by a male bird, then typically in less than a month, in the right conditions, that egg will have somehow produced a chick.

How is this possible? How does nothing more than protein and fat and some minerals amazingly and reliably form a chicken, or any bird for that matter? With mammals, the fact that they are physically attached to the mother via an umbilical cord makes it all seem more plausible. But with a chicken, that egg is completely independent of its mother. And given the right temperature and conditions it will breathe and hatch without the mother anywhere near it. It is a truly unique and amazing thing to ponder.

Similarities In Embryos

Setting aside the hocus-pocus poof! of a magician god, it is likely that things formed in a somewhat normal manner after the "let the waters bring forth" event occurred. So it is the specific "let the waters bring forth" event that we are focusing on at this point.

Evolutionists look at similarities of organisms and make claims based upon these similarities and assert that they share the same original single-source lineage. When looking at many different types of creature embryos, there are striking similarities in the very early stages. So, does this then mean that they are related through a blood line? Not necessarily. Similarities are certainly evidence of something, but are they evidence of lineage, or are they evidence of design? And this is where the divergence between people's perception of the Biblical account of creation and pop-science takes place.

I would like to present for consideration an example that demonstrates no other option than shared attribute regardless of a Creator creating or evolution occurring. And the example is best presented as "resolution". Consider the resolution of anything. If you watched a one pixel resolution TV then everything would appear identical other than possibly color. If the TV was two pixels then you might be able to differentiate between two pixels depending upon what is being viewed.

Of course this is only a simple illustration, but with life, the cell/egg is fertilized and the initial single fertilized cell (pixel) will appear mostly identical in any breeding creature. This holds true at least up until the cellular resolution becomes vast enough to display any recognizable forms. But this presents another problem in determining creature types. To the best of our scientific knowledge there are certain requirements for breathing creatures to form, and they tend to form in generally the same order, with heart, brain, and spine forming very early on. Now since the cellular resolution is quite low at this point in the gestation process, a heart, a spine, and a brain will all appear generally the same until the cellular resolution of each organ becomes high enough for us to be able to determine its type. This same thing is true for the other key parts of the creatures' form, such as the limbs and gender-anatomy etc. Please keep in mind that all of this is referring to the very early stages of development of nearly any creature.

The point being made here is that there is no choice for things to appear similar when the cellular resolution is yet very low in the developing stages, and so to claim that these similarities prove shared lineage is a major scientific blunder. To add to this a bit more, the Biblical account of creature arrival is broken into the particular type(s) brought forth on days five and then some additional types on day six, but within each day there are further categories.

On day five we have:
- The creeping creature having life.
- The winged fowl that may fly over the earth.
- The great whales.
- Every living and moving creature.

With each of these very broad groups, a substantial increase in cellular resolution is required to determine any divergence from one type within the group to the next type within each group.

It is here that we have to see whether or not it is at all possible to somehow marry the Biblical account with the scientific account. And this is also the point at which a major divide between science and the Bible often occurs. Evolution proponents claim that everything is descended from a common ancestor. If this is true then evolution is obviously spot-on. But here we are forced to question these similarities and wonder if they are similar due to descendancy, or if they are similar because of design.

Brachiopods

What are "Brachiopods"? Sounds like a cool name, but what exactly is it? To the untrained eye if you were to encounter one of these ancient creatures you would immediately assume that it is a random breed of a typical clam. But according to "science" you would be utterly wrong! Brachiopods are said to be somewhere between two-hundred-fifty-million and four-hundred-fifty-million years old, and supposedly no longer exist. They are different from clams in many ways, much like Asians have different characteristics than Africans and Europeans etc., yet they have many more similarities. On one side of the debate, brachiopods are indeed ancient clams and, do certainly to the best of our scientific understanding, prove that evolution does occur to a minor extent. According to science and the fossil record, these brachiopod predecessors of the modern-day clam have evolved into the modern-day clam, but the divergence is very small considering their claim that a passage of somewhere between two-hundred-fifty-million and four-hundred-fifty-million years has transpired. That is quite honestly impressive stability or similarity of form being displayed over a very long period of time. That level of stability could cause one to question the proposed notion of evolutionary deviation of any one kind.

Archaeopteryx the Dinosaur

Getting back to the chicken or the egg subject, let's consider what is believed to be the first bird or the proof of common descendancy of dinosaurs and birds. Enter Archaeopteryx! It's a bird, it's a dinosaur... it's a chicken?

Archaeopteryx is a fossil found in Europe and is believed to be about one-hundred-million years old. This "dinosaur-bird" must have been massive or freighting in size since dinosaur means terrible-lizard. Measuring in at a whopping eighteen-foot wingspan... No, wait a minute! That's an eighteen-*inch* wing span? That's right, this "terrible-lizard" bird or dinosaur bird as it is often referred to in pop-science, named "Archaeopteryx", is the size of a large crow or pigeon, and has all the markings of being what most everyone one Earth would refer to as "a bird".

Calling "Archaeopteryx" a dinosaur is quite honestly a dishonest way to promote the fossil. But scientifically it does clearly prove that bird-like winged creatures did exist a very long time ago, one hundred and fifty million years by their estimate of the age of the fossil, and now here again we have incredible stability of basic creature form over vast amounts of time. Archaeopteryx proves beyond a doubt that birds are indeed very old and over the claimed one-hundred and fifty million years have changed very little considering the claims of evolutionary deviation that should have occurred during that period. Of course, this is not to say that other creatures did not descend from Archaeopteryx, but that evidence is scant. If you get the opportunity, take look around for pictures of Archaeopteryx to see the various artists' renderings of what each artist imagined in their own mind what this pigeon-size "dinosaur-bird" looked like. There are more than a few options for you to peruse. Which do you think is the right match if any? Does it look like a dinosaur? Does it look like a bird?

Chapter 3

Let the Earth Bring Forth

In the last chapter it was mentioned that there are two distinct creation events for creatures, that of day five and that of day six. Day six is the "Let the Earth bring forth" group consisting of:

- The living creature in its kind.
- Cattle
- And creeping things.
- Beasts of the earth, according to their kinds.

When the "waters" brought forth, it appears to have been mostly sea creatures and birds and some creeping things, but when the Earth brought forth it appears to be land creatures that do not fly: beasts, cattle, general creatures and creeping things. The text does appear to infer that the creatures brought forth by the Earth were basically land creatures. This possibly indicates that the creeping things that the water brought forth were mostly amphibious in nature, and the creeping things that the Earth brought forth were more than likely not water dwelling creatures. This is consistent with what we experience in the real world as we live this very day.

Elevate Your Thinking

Trying to establish baseline information is difficult on a tumultuous globe such as our Earth where the land surface changes over time. It would be easy if things made sense, but logically speaking, we find things in peculiar places. Since there have been fish fossils found at some relatively high-altitude locations we must consider how those fossils may have gotten there.

This is another area of great disparity between the Biblical accounts of Earth and the scientific accounts of Earth. Any sea creature fossils found above sea level makes us wonder a bit. Now if those fossils are found in low lying areas it is no stretch to imagine that the water levels were at some point raised twenty or thirty feet, thus allowing the fish or sea creatures to have access to such areas and die there. However, the story is entirely different if those sea creatures are found in a mountainous region at high altitudes, which they have been.

Here we have a few options to make various cases. The first is that the entire globe flooded, which is how the fish fossils that have been found at high altitudes got there. Then there is the long-age theory suggesting that the fossils were formed at the base of an ancient lake and over time the upheaval of the land caused the lake bottom to rise up to mountainous levels, thus placing the fossil found today at high altitudes. But there is also another potential explanation where a waterspout over the ocean/sea could have drawn out small sea creatures and deposited them at higher altitudes. And finally, we have the simplest explanation which is that a predatory bird such as a large eagle plucked a fish out of the sea and dropped it on a high location, and the fossils found are from that sort of event. Now determining which if any of these options is the proper explanation should be able to easily be determined by assessing the immediate area and material surrounding the found fossils.

If the sea creatures were dropped by a predatory bird then they would stand alone and would be very few in number with no other logically accompanying sea creature fossils nearby. If the sea creatures were deposited as a result of a waterspout then we might find several other types of somewhat random sea creatures in the vicinity of the primary fossil find, but no other organic sea materials. In regard to a global flood, we would expect the depositing of some other organic material and other sea creatures in the area, but it would be very limited since the global flood is said to have been a relatively short period. And lastly, if it was an ancient lake bed from low lying area that was lifted up over many years, then we could logically expect to see many fossils of other large and very small sea creatures and other organic matter fossils found in the same area; in fact, immediately around any found fish fossil it would be abundant in that case.

If a fossil was naturally deposited anywhere above one-thousand feet above sea level and not showing any evidence of land upheaval it would testify to a massive worldwide flooding event regardless of whether or not such a flood was able to completely cover every mountain top peak. Knowing such details of the surrounding material of any high-altitude fossils found is what should determine our final conclusion about a particular find. There is more on this topic in the book *The Science Of God Volume 5 - Boats, Floods, and Noah - The Deluge*.

Animals On High Ground

The previous section explains the various ways sea creatures could be deposited on what today is higher ground. Each possibility offered is plausible, and in various cases each scenario has likely occurred somewhere on this earth. The first part of the sixth day where the land animals/creatures were created, these creatures would likely be those that we see thrive on land but indulge themselves in water occasionally. We cannot easily define these in reptilian versus mammal groups between the day five and day six groupings, but the day five groupings tend to be egg-laying creatures much more so, with potentially a small variety of mammals such as whales. Where the day six creatures tend to be without external eggs and carry their young in the womb until birth. The general mammal and reptile/bird groupings are strong, yet are possibly breeched on both days. However, we must be careful in making such assessments because we currently categorize these creatures as we do and because of this we are mentally influenced to see them in our own modern categorical eyes potentially causing us to be missing a more logical categorization that Biblically speaking would distinctly classify them into completely logical groupings.

Defining Animals

"Animals" move, which has the same root as "animation". The word animal is derived from the word "anima" meaning *breath* or *soul* or to *breathe* or *spirit*. This implies that free-air breathing is important to "animals".

"Mammals" are creatures with mammary glands that produce milk for their offspring. Most of the creatures brought forth on day six fall into both the "animal" and "mammal" categories.

"Reptile" is a word that is derived from Latin *reptus*, meaning to creep or crawl.

If you consider the groups mentioned for day five and then those for day six, the general divisions make sense between egg-laying creatures and general mammals, although there are those crossover creatures like whales mentioned on day five. Day six conceivably describes only non-egg-laying creatures.

If the day five and day six separation of creatures did have anything to do with day five creatures-**with**-eggs versus day six creatures-**without**-eggs, it would indicate that what we refer to as dinosaurs would have been brought forth on day five by the "waters". Fossil evidence strongly indicates that many dinosaurs lay eggs. This is also consistent with science in that science tends to group dinosaurs and birds together due to certain similarities.

This is an area where both science and creation must be cautious not to ignore good information. Those who study creation tend to discount much of what evolutionists have uncovered during their archeological digs regarding similarities in creatures. And the evolutionists tend to ignore the entirety of the Genesis creation text. Neither approach is helpful to getting to the truth of how it all occurred regardless of whether or not it was at God's hand.

Chapter 4

Making a Way Using Scientific Extrapolation

Scientific extrapolation is one of the most useful scientific tools and it is also the single most damaging scientific tool when it is abused. Without scientific extrapolation there is little that would be accomplished in this world. Luckily for all of us, engineers are much more proficient at using extrapolation than is science. "Science" has the luxury to imagine and espouse the unprovable, where engineering on the other hand does not. If engineers fail in their extrapolations then someone is likely to be injured, or worse, killed in an accident resulting from inaccurate extrapolation, and/or the erred engineer will be fired. Science can make up anything they want as long as it does not have to be proven in real-world settings. Something such as the extrapolations that are done by rocket scientists, or better described as engineers, must be accurate, provable, and must work or an astronaut will die. But when it comes to the origins topic and evolution, the extrapolations seldom interact with critical real-world situations, thus evolutionists can make up any explanation that seems plausible, and seldom are they required to prove any of it. Those are very low "scientific" standards.

Extrapolating Finches to Crocodiles

While watching some "documentaries" about some very imaginative extrapolation, one of the evolutionists being interviewed discussed a connection between finches and crocodiles. Now if we are going to allow this level of evolutionary extrapolation, then we can connect finches and crocodiles together or to anything else for that matter.

The logic discussed earlier that DNA cannot be added to is in direct conflict with the very notion of evolution and evolving from a single cell. When an appendage evolves out, we have to question if that is a subtraction from DNA or an addition to DNA. While an appendage being phased out seems to be a deduction, what we don't know is if that appendage's DNA instruction changed in some way to an alternate purpose or if that appendage is in the DNA but is somehow for some reason blocked from developing.

If you partake in discussion of evolution and begin to ask the tough questions, you will quickly find that it is a convoluted pile of theories that have no central point of reference other than Darwin. This is best seen in any attempt to define the word "species" for scientific purposes.

This might make it seem like evolution is out of the question and that the only possibility is the Biblical account, but this is simply not true.

Evolution Fails the Scientific Method

When you take a deep dive into the evolution-versus-creation debate you tend to hear a great deal about the "scientific method". When looking into the scientific method you then tend to hear terms like *objectively demonstrable, repeatable, observable, verifiable, falsifiable, testable facts*, etc. Darwinian evolution typically fails every one of these categories. Oh yes that statement will be vehemently argued by every evolutionist, but

Darwinian evolution, specifically, is mostly speculation and wild extrapolation. However, evolution does exist and it is actually very rapid. Darwinian, or single-source long-age, evolution does not meet the criteria for the "scientific method".

Whether or not something is "falsifiable" is used as an indicator as to whether or not it is to be considered "scientific". Theories are falsifiable when they can be contradicted by observation. When observation is impossible, falsifiability is not possible. Thus, the claim that there is a "God" is not *scientifically* provable or disprovable as far as we know, so the claim that God exists is not specifically "scientific". Things are somewhat different with Darwinian single-source long-age evolution. Single-source long-age evolution is simply *not* observable because it occurred before we were here and alive to see it, thus we simply cannot observe it which is similar to the God issue. But differing from the existence of God, evolution has some wildly creative, some erred, and some outright dishonest extrapolation that is commonly used in the standard pop-science evolution models.

Is It Really Demonstrable?

What does "demonstrable" mean? And is single-source evolution "demonstrable"? Is it really demonstrable that all creatures have a common ancestor? I had the opportunity to partake in various discussions on this very subject and found many people on the evolution side of the debates to get quite hostile. In these discussions, it was not the two sides opposing one another, which is to say creation versus single-source evolution, but rather simply asking questions of evolutionists. I was surprised to discover the animosity projected by evolutionists if you questioned their claims stating that evolution is "demonstrable". Many claims are set forth that long-age single-source evolution is demonstrable, yet no specific linkage is able to be produced. Now, of course any readers who support evolution will at this point be upset by that statement, however, when looking at the fossil record there are major gaps in the forms. And the connecting of forms that is often done uses a great deal of imaginative extrapolation and artistry.

You might then wonder where these extrapolated connections come from. To that I would ask, where do all of the pictures of creatures come from that are shown in the evolutionary trees of life that we see? Which of any of the pictures with all of their forms, scales, feathers, colors, etc. do we have actual verifiable, falsifiable, evidence of?

These questions tend to invoke the "forensic" argument regarding human remains that that police investigators have done that are not always, but very often, based upon photos or other family members. These models are built upon the skulls of the decayed bodies found and are based on our understanding of the human structure. These same principles are applied to the creature realm and they create artist renderings of creatures based upon what? Other dinosaurs alive today? These depictions are generally derived from lizards but more often they are from preconceptions the particular artist had from seeing other artists' conceptions of other similar dinosaur remains in those previous artists' drawings. It is all made up artistry based on only speculation.

This section might come across as a condemnation of evolution, but it is not. It is a condemnation of excessive evolutionary extrapolations that are *not* falsifiable or verifiable and are typically promoted by pop-science as "undeniable fact".

The Paper Trail a Mile High

As humans, there is a yearning in us to understand our origins and the origins of most everything that exists. In our attempts to do so, we try to package everything into little compartments that make sense to us. Once we have formed those little boxes of mental data, we have a tendency to not allow those boxes to be opened or examined by anyone who might challenge us, and we often react viciously when someone attempts to pry our boxes open to verify the data in any of those little mental boxes in our head.

There is a great deal that we do not know, so to claim "undeniable fact" in evolutionary science, if single source evolution can technically be called "science", is a stretch to say the least. Let us liken the time-spans involved to a measurable visual to get a grasp on our life experience compared to the age numbers used when describing evolution.

An example used in previous volumes of *The Science of God* illustrates time in a visual way picturing time in years as distance and as a single stack of paper standing upright with one piece of paper lying flat upon the next paper. Each piece being only about .004 (4/1000) of an inch thick, which is about the thickness of a typical piece of paper you might put in your printer on your desk, or a piece of "typewriter" paper if you are older. When doing the math you will find that at this thickness of paper that one inch of paper on the stack will be 250 sheets. One foot will be 3000 sheets, and a stack a mile-high will have 15,840,000 sheets. So as to not miss any zeros that's nearly sixteen million sheets in the stack. A billion sheets would be a stack just over 63 miles tall.

In effort to keep our stack mentally within view let's make each sheet equal to one thousand years. This means that the mile high stack would roughly represent the claimed age of the Universe. Now, since Earth is said to be about three-and-a-half-billion to four-and-a-half-billion years old that means that a stack of paper a quarter of a mile high or roughly 1300 feet high (a skyscraper's height in paper) is the age of the Earth with each paper being one thousand years.

Now since evolution is said to have begun about four billion years ago, your life experience is about the thickness of a tenth of a piece of paper. That is the thickness of a single layer of typical kitchen plastic wrap you would use to wrap a sandwich with.

Compare that to the one thousand three-hundred-foot stack of paper.

Now there are thousands of fossils of creatures that are said to prove descendancy from a single source over a period of billions, and at minimum hundreds of millions of years. Do we truly imagine that our scientific partial thickness of paper of say maybe 500 years, or a half sheet's thickness, is enough research to "undeniably" conclude that single-source evolution *is* the way it all occurred? That is highly unlikely. There is a great deal more research for science to partake in before we will be able to "undeniably" prove anything of the sort and call it "fact".

Now in this context, I mention this paper stack also in relation to radiometric dating methods often used to date fossils. Since science, as we tend to think of it today, is relatively new, less than five hundred years at the writing of this book, is it a reliable method to make adamant claims of hundreds of millions or even billions of years of age for any fossil? We can't even accurately age, in a blind-test, the fallout from the Mount St. Helens volcano eruption near the end of the twentieth century of which we know the exact age of because many people witnessed it. How then do we imagine we have an accurate estimate of the true age of fossils? No self-respecting scientist should insist on an *exact* age of a fossil, and since age results of modern volcano eruptions are several-thousand-times-over *inaccurately* estimated via radiometric dating, then what of hundreds of millions of years? That could make a hundred million years very far from accurate to the tune of that actual age being as little as fifty thousand to one hundred thousand years. That is a very big difference.

The point here is not to diminish radiometric dating, but rather the abuse of thereof, as well as the abusers not informing people of the reality of the vast extrapolations used in concocting the evolutionary tales that are presented as "undeniable" when using dating methods that allow such vast variance.

Evolutionary Laws?

"Laws" is a strong word to be used in conjunction with something that is not particularly falsifiable. But, before we get too far down this rabbit hole, we must differentiate provable biological evolution that actually can be observed before our very own eyes within our very own lifetime, apart from that which cannot, such as single-source Darwinian-style evolution.

In bio-cellular evolution we can see it, witness it, verify it, feel it, and touch it all while it is occurring. The same is simply not true of long-age evolution that extrapolated every single aspect of its theories beyond what is reasonable. The "laws" for biological evolution can be established through repeated experimentation. But this is *not* so with single-source long-age evolution. We have no way to experience a hundred million years to see the changes that *theoretically* would occur. However, we can see changes in a single generation of crossbreeding various varieties of dogs for instance. Any claim of evolutionary "laws" regarding long-age single-source evolution is simply dishonest—it is a lie.

Chapter 5

At First Breath

Most creatures breathe in some manner, but land mammals all seem to share a similar method of general nose and/or mouth breathing. These similarities in fundamental construction are either likely due to evolution or to specific design. If all creatures evolved from a single source, then the similarity of the functions of body organs would certainly make sense, but that view leaves a great deal to chance. On the other hand, creatures being specifically designed allows for simple and logical explanation of similarity in design that takes into account the otherwise largely unexplainable aspects of each body organ.

If a system is created to use blood to distribute nutrients throughout the body of a creature in order to supply each organ its needed nutrition, then we would expect that a similar use would be applied in any of the creatures that were part of any broad creature group. And since blood needs to somehow be circulated through the system, some sort of mechanism to force that blood through the system would be required. This does not allow for a whole lot of room for methods. While there are more methods than only hearts to pump the blood found in most creatures, the general heart pump is a good robust design that seems to have a great amount of versatility across a wide spectrum of creatures.

That is to say that when something works, why invent a new method for each implemented type? Instead, repurpose the design for many similar uses. This can be likened to when we create something such as an engine and then use those engines that are all generally similar in design for automobiles, trucks, vans, cranes, motorcycles, go-karts, generators, mopeds, toy airplanes, and so much more. Because the design is well done and versatile we can review it and say that it is good, and then we can go forward using it wherever such power is needed. Assuming that a Creator exists, we must consider this line of rationale. If something is specifically designed, as opposed to having been able to randomly evolve, we would *expect* that **systems** would be developed and used repeatedly on creatures sharing a similar nature of design.

Thus, if a creature uses the *blood distribution and exchange of nutrients for impurities model*, then other systems required to supply the nutrients and filter out any impurities would also be required of most of those creatures that share that fundamental system design. In other words, we would then generally expect kidneys, veins, arteries, bladders, intestines, stomach, throat, mouth etc. or some thing or method that deals with those needs. These are all verifiable and falsifiable systems and almost all of them are shared throughout the entire animal kingdom in nearly every creature.

Interpretation Manipulation

Now at this point we get into interpretation of data/findings. Data and findings can be two different things. Something that is "found" is something that you have seen, touched, and witnessed firsthand. You can then take the description of that finding and enter it as data for other people to consider. This is where most of our problems and debate enter the scene.

Few people catch the difference between *actual findings* and *scientific conclusions*. If you find a diamond ring lying on the ground, then you found a diamond ring lying on the ground. You can weigh the diamond, and describe its cuts and determine the gold or silver content of the band of the ring. This can be done to a point where any person who reads the recorded data about the finding can clearly picture, in their mind, almost perfectly how the ring looks. This is repeatable, and if the ring is retained it can be further examined by others to verify its makeup and characteristics or attributes and the original analysis.

Now if the discoverer of that ring begins to speculate as to the origin of that ring; meaning, who lost it and why or how it was lost, that possibly will not be provable or verifiable. So let us imagine that the previous possessor of that diamond ring has long since passed away and that the ring has no inscription in it. Now if the recent finder of that ring begins to speculate that a jilted lover got angry and tossed the ring out the window, is that undeniable fact? Or is it speculation?

In this case it is absolute speculation. The finder is interpreting the ring found as having been cast away as a form of rejection to the said jilter. But is this accurate? Since in our example the previous possessor of the ring has long since passed away, and we have no inscription in the ring and no idea who owned it, is the interpretation of the data falsifiable, verifiable, or provable in any way? The obvious answer is, no. Could the speculation be right? Yes, but is it right? The answer is that we simply do not know.

This example lines up perfectly with the imaginative speculative abuse of "undeniable" single-source long-age evolution, but let us not forget that it is little different with the six-twenty-four-hour-day creation theories that abound in many Biblical circles. Is either theory-set verifiable or falsifiable?

When discussing this with six-twenty-four-hour-day creationists they very quickly run into trouble when forced to answer tough questions regarding the "fossil record". While neither side of the debate can truly give answers that are "undeniable", it seems most disingenuous for a six-twenty-four-hour-day creation believer to suggest that the Creator of all things would "make the fossils to look old". This is a foolish and simply inconsistent rationale for a Creator, that is to say a Creator that would have to exhibit tremendous logic and rationale in order to develop highly versatile and complex systems for the creatures that were created.

But there is another side to the interpretation of things. In this I am speaking of the scientific view of the Biblical account of creation. Now this goes back to the beginning of this book discussing the various Bible versions that we all now have access to and our need to use versions that are more authoritative which date back many hundreds of years.

Shortly after the Luther Reformation Revolt, there was a fervor to reinterpret the Bible and to make it more understandable to the common man in their own language. Previously, obtaining a copy of a Bible was cost-prohibitive for the common man, and it is likely that most people could not understand the language that the Bible was available in, namely Latin. So, well-meaning people worked to translate the Bible into their native languages such as English and German. This is all good and in general was quite accurately done, until the printing press became highly automated through reciprocal mechanization and movable type, thus greatly reducing the cost of replication and prompt distribution well after the introduction of the Gutenberg press. The real problem occurred in the seventeen- and the eighteen-hundreds. Many of the newer translated/transcribed Bible versions created during and after that time had greatly altered the Genesis One creation accounts to reflect the interpreter's own interpretation of what occurred during the creation events of Genesis One. This is obviously a somewhat unavoidable situation because we explain things as we understand them. But in many cases these post-Reformation versions, are nothing more than a *rewording* of the Bible version that was already translated into their native language. In other words, many of them took an English translation and reworded it in English *to better suit their own understanding* of the text. Nowhere in the Bible is this more evident than in Genesis One's creation account.

This might seem somewhat insignificant until you realize that many people have been brought up having been taught from Bibles with highly inaccurate interpretations of the creation events. And since they were taught that "The Bible is infallible", all Bibles, to them, are created equal. This is the general assumption by most everyone who is not specifically studying the Bible. That is the purpose of the book *Understanding The Bible – The Bible How-To Manual* AND *The Things We Don't See*, which points out many things we don't typically know, see, or notice about the Bible, such as the types of translation inaccuracies being discussed here.

The point here this is that when you have a bad translation of the Genesis One text you are immediately disadvantaged and immediately have an inaccurate view of the text. This is especially true regarding the First Four Days of creation as is discussed in *The Science Of God Volume 1 – The First Four Days*.

If a person has been brought up using a Bible version with inaccurate text, then they are far more likely to adopt an inaccurate six-twenty-four-hour-day creation model. This is fine for them and generally will not affect their "salvation" due to their innocence in the matter. But the problem comes in when they enter college and are challenged on their interpretation of the inaccurate Genesis One creation account that they have been given, taught, and indoctrinated with. Most people do not even know or realize that these erred reinterpretations of the Bible even occurred. A challenge to their beliefs is only the tip of the interpretation-manipulation-iceberg. Many of these discouraged believers ultimately reject the Bible entirely and as a result, they end up rejecting God, and then many go on to become teachers and thus spread their Biblical misunderstanding and rejection of the Bible to other unwitting impressionable students.

Now because they are only familiar with the poorly translated versions of the Genesis text and are generally not aware that other more accurate versions exist, they will vehemently insist that challengers of their newly adopted godless "The-bible-is-just-fantasy-stories" beliefs about the Bible must accept that the Bible is fantasy, when in fact it is not. If you are curious about the differences in Bible versions and the problems they cause, I highly recommend taking a look at *The Science Of God Volume 1 – The First Four Days*, and the book *Understanding The Bible – The Bible how-To Manual* AND *the Things We Don't See*.

What I am getting at here is that many people who have adopted the single-source evolution model do so due to these perverted post-Reformation versions of the Genesis One creation text. It hurts to have your beliefs beat down in a public forum such as a college classroom. When a student is certain in their own mind that they have it all figured out after having been indoctrinated from youth on with some of these post-Reformation Bibles, they are essentially ill-equipped to debate their science professor in a public forum and will essentially be beat down and publicly humiliated in front of their classmates or other peers. These public beat-downs would be non-existent if we would all simply use logic.

Analysis of Terms

Terms are a truly abused aspect of the evolutionary sciences, most notoriously the Darwinian type evolution from one kind of animal to another. The first time I read Darwin's books I was left with the question of what specifically defines a "species". For instance, we look at humans as a "species" distinct from apes or monkeys. That's fair enough. But if you look at the differences between "species" of finches or between "species" of beetles as described by Darwin, then humans certainly have different species within humanity. The extremes of the various human skin colors are enough to classify "species" of humans when using the same level of categorization that was used by Darwin and his followers to this day, yet we do not do that. The point being driven home here is that science is sometimes not very scientific and is very inconsistent in categorization, and nowhere is this more apparent than in the creation subjects. In this case it is very specifically the *definition* of "species". Who gets to determine where a "species" begins and where it ends? This is where the evolution trail gets very convoluted.

Define Terms

An age-old debate regarding creation centers on the definition of "kinds". How do you define a "kind" as described in the Genesis creation account? This is where the debates can get very heated and is where evolution comes in with great force. The Bible is very limited in its description of "kinds". In fact, there is no definition of "kind" in the Bible, and in Latin versions, the specific word used is "species". "Species" is not a new scientific word as we inadvertently assume, it is an old word that has been translated into English as "kind". Thus if debates arise demanding a definition of "kind" it is the same as the definition for "species" in the case of the Genesis One creation text. There is no real scientific definition of either "kind" or more specifically, "species".

Douay Rheims English Genesis 1:25

"And God made the beasts of the earth according to their kinds, and cattle, and every thing that creepeth on the earth after its kind. And God saw that it was good."

Latin Vulgate Genesis 1:25

"Et fecit Deus bestias terrae juxta species suas, et jumenta, et omne reptile terrae in genere suo. Et vidit Deus quod esset bonum"

When looking into this you will quickly find tremendous disagreement, scientifically, regarding what the term "species" means. You will find many definitions, mostly having to do with the particular evolutionary science you are looking into, but even within that there will be disagreements on the specific definition of the term "species".

The debates about the definition of kinds/species are all about the splitting of scientific definition hairs, which the Bible does not do in the Creation accounts. The problem comes in when supporters of the evolutionary theories challenge creation supporters and insist that the creation supporter define "kind". Typically, the creation supporter will take the bait and proceed to attempt to define "kind", basically not realizing that "kind" is merely a Latin-to-English translation of the Latin word "species". Of course, their definition is typically delivered with many inconsistencies and met with many questions, so then they typically ask the evolution supporter for a definition of "species" and the evolution supporter will rattle off a list of attributes of a "species". But when pressed on the issue, the evolution supporters will also begin to stumble. When you look into this definition issue, you will see that the creation supporters, whether right or wrong, are typically more consistent than the evolution supporters are in their definitions.

To remedy this, evolution supporters have adopted terminology that solves this problem... sort of. This is where terms like *genus*, *species*, *clade*, *monophyletic*, etc. enter the scene.

"*Genus*", is class, kind, or group with common characteristics or with a single common characteristic.

"*Species*", are living organisms of similar creature capable of interbreeding.

"*Clade*" has a common ancestor and includes all of its descendants.

"*Monophyletic*" is a group of organisms classified in the same taxon and share a recent ancestor.

But we'll stop there because things can go on for a while because there are more, and the deeper you crawl into that rabbit hole the more convoluted it becomes. The problem is not with the language; rather the problem is with the decision of classification. *Who* specifically gets to decide where the line is drawn on the many fossils found? And more importantly, who gets to decide where the various artists' renderings of the creature belong in the classification process?

The problem with long-age evolution is that there are no hard facts, and everything is little more than guesswork, it is all speculation presented as "undeniable fact". It is neither undeniable nor factual, though that point will be hotly debated by any evolution supporter. We often see a bait-and-switch being pulled when evolution is discussed, so it is critical to define what is being discussed at the inception of a discussion about evolution. There is an enormous difference in evidence between *biological* evolution and *single-source long-age* evolution. What often occurs, is that when you question-corner a Darwinian or single-source long-age evolution supporter, they tend to use the bait-and-switch routine by invoking "biological evolution". While it is true that you would need biological evolution in order to achieve single-source long-age evolution, there truly is no legitimate comparison of the two. Biological evolution can be witnessed in a petri-dish in a lab and actually viewed with a microscope, where single-source and/or long-age evolution, on the other hand, cannot.

The definition problem is vast in the evolution-versus-creation areas of discussion, and both sides of the debate have issues to deal within the definitions areas, though admittedly worse on the evolution side of the debate. The evolution side tends to add new terminology on occasion to explain away inconsistencies. Ultimately both sides are trying to understand the same thing and generally do so by trying to create little mentally comprehendible packages. Sadly, while doing so, pertinent data from the opposing side is typically ignored or outright rejected only because it is from the opposition. If only there were scientific dictionaries.

Does a Science Dictionary Exist?

The topic of "definitions" quickly gets very messy. When you begin to ask deeper tough questions of either side of this debate you will quickly find that, for the most part, very few people have any concise answers to offer. Initially, the conversations are filled with many words that are compelling, but when the chain of subsequent questioning continues, things change quickly as the person offering answers is no longer able to adequately give answers. This is neither good nor bad, it is simply a reality of life because no one knows everything, and eventually we will exhaust someone's knowledge when we ask them too many questions. This usually causes them some frustration that often will come out as anger towards you and your questions.

I have asked around for and looked for a dictionary that offered definitions of the many scientific evolution terms that have been used in such discussions, but was not able to find one. Though there are many books written on evolution, they all encounter the same problem–they use language that is not particularly clearly defined. While some terms are clearly defined here or there, they are often used in a willy-nilly manner, but that is mostly done by armchair evolutionists. The same is true of the creation side of the argument. With the creation side, it is more or less a problem of Bible version selection that presents their problems.

Even if we were to create a specific science evolution dictionary, we would still be faced with things like the many definitions of "species", and thus the potential misuse of the term in any given situation. Being especially true in the world of science, when we define things in a rigid manner, it inevitably causes the need to sub-define, and thus introduce newer refined terms to describe the newly noticed distinctions. This in itself is not a problem, but what does become a problem is that we trap ourselves in these fancy little mental language boxes and then can no longer see the broader topic. We end up not being able to see the forest through the trees.

Chapter 6

Two of a Kind Makes a Full House

Outside of peculiar and quite rare examples, two of a kind ultimately makes the "kind" complete. The male-female model is undeniable in all of the animal kingdom. Despite what modern culture might choose to invent in their minds in this regard, there is male anatomy and female anatomy, and without that anatomy and the functionality thereof the "kind" could not exist. This is undeniably proven with each successive, testable, verifiable, falsifiable, witnessable generation of creature. Without the affinity of male to female within the "kind", there would be no kinds. Even if a Creator did somehow start the DNA ball rolling, without the largely unexplainable attraction within kinds, each "kind" would disappear very quickly!

Kinds

Kinds or Species, Species or Kinds, it is all the same. From a Biblical translation perspective, we must consider the two as the same thing. Here it is again for your consideration:

English Douay Genesis 1:25

"And God made the beasts of the earth according to their **kinds**, and cattle, and every thing that creepeth on the earth after its **kind**. And God saw that it was good."

Latin Vulgate Genesis 1:25

"Et fecit Deus bestias terrae juxta **species** suas, et jumenta, et omne reptile terrae in **genere** suo. Et vidit Deus quod esset bonum"

This takes us back to the groupings between and within the days five and six of creation. What are the specific divisions and what has that to do with "kinds"? As just mentioned, for our purposes here "kinds" and "species" are the same. So the distinction we are trying to make is what specifically separates "kind/species"? Since this book is mostly analyzing an English version of the Bible we will generally be using the term "kind", but always keep in mind that whenever it is mentioned we could just as accurately have used the term "species" from a Biblical interpretation standpoint.

Biblically speaking, what determines "kind" in the Bible? And how many kinds are specifically mentioned? This is the area where many creation supporters trap themselves. The creation account is very basic and likely so for at least one reason in particular: very few people read the Bible—it's rare that people even complete Genesis. And sadly, there are very many people who have never even read the brief Genesis creation account which is only about five minutes of reading. Thus, having a lengthy detailed explanation of the deeper details of the *how* and *why* of creation would without question fall upon deaf ears. A detailed account of creation would likely fill many volumes, with each volume being far thicker than the entire Bible itself. So if the Bible's Genesis One is a true account of creation, then relaying to us only the basics was the wise choice.

Yet in the brief creation account given in Genesis there is a truly vast amount of information that is typically overlooked.

Instead of properly scrutinizing the Genesis One text, creation supporters often invent or make assumptions that step far outside of the actual words given in the text, and in doing so trap themselves in their own words. When questioned, they will work to define things and are confused by *species* versus *kinds*, and then fail to understand that, Biblically speaking, "kinds" and "species" are the same word. This causes them to become tangled in modern scientific terminology definitions placing them in the hands of their often-hostile debate opponents.

Douay English Genesis One:

"[20] God also said: Let the waters bring forth the creeping creature having life, and the fowl that may fly over the earth under the firmament of heaven. [21] And God created the great whales, and every living and moving creature, which the waters brought forth, according to their kinds, and every winged fowl according to its kind. And God saw that it was good. [22] And he blessed them, saying: Increase and multiply, and fill the waters of the sea: and let the birds be multiplied upon the earth. [23] And the evening and morning were the fifth day. [24] And God said: Let the earth bring forth the living creature in its kind, cattle and creeping things, and beasts of the earth, according to their kinds. And it was so done. [25] And God made the beasts of the earth according to their kinds, and cattle, and everything that creepeth on the earth after its kind. And God saw that it was good"

How many types can you count in the preceding text? Do you see dogs, or cats, or any other superficially named creature? Or are they very broad groupings? It is important to note that the translation of the Latin "Creavitque Deus cete grandia" to "And God created the great whales" is potentially creative translation in itself. The Latin text indicates essentially large or tall creatures rather than specifically "whales", as is indicated in some English translations, or "sea monsters" as is seen in other English translations. In this case the Latin text being translated to either "whales" or "sea monsters" is potentially taking some liberties with the text. Most English versions use one of the following "whale", "sea creature" or "sea monster" and maybe more accurately one translation uses "great dragons". But in all of these it takes a bit of liberty with the text. The implication is generally that of large water creatures in these translations. This assumption makes some sense since it is the "waters" that brought forth. But since the "waters" also appear to have brought forth the birds of the air, we cannot automatically assume that it implies specifically that the particular creatures in question are water creatures. It could just as easily be what we commonly call "dinosaurs".

As you can see, the division of "kinds" is really somewhat of an obscure topic regarding the definition thereof in the Bible's account of creation. As mentioned in earlier Volumes of *The Science of God*, Genesis One is unique in nature, in that it cannot be taken as anything other than an account of the creation process which could not have been witnessed by mankind according to the order of events. This means that a great deal of our translations and interpretations of the text are influenced by our post-creation understanding of things, which is again conceptually a chicken-versus-egg issue. In other words, what came first, the concepts or the words? Meaning how do you refer to something before it is a thing? How do you refer to an area where a city now sits, but before the city was ever imagined to be? Typically, we might simply say the city name and refer to it as "before the city was there." Use this same line of thinking regarding all of the text in Genesis One. This also applies to the very limited listing of "kinds".

Why Versus How

As we venture down the creation path, we get into many areas where making divisions is an important task. In the previous Volumes of *The Science Of God*, separation or divisions are also discussed. The importance of these divisions needs to be recognized by anyone who wants to be remotely serious about trying to understand the creation account in Genesis. A key mental division we must become acutely aware of is the difference between *how* and *why*.

The implication of *why* is purpose. The implication of *how* is function. We can examine *how* an engine works and describe it in deep detail. Often, children will ask *why* when seeking *how* answers, which can be somewhat frustrating to the parents, where if the child asked *how* then it would not seem as frustrating because the *how* answer is straight forward, where *why* involves intellect. Asking "*Why* did you do that?" demands a reason, *how* does not. So while these two, *how* and *why*, might at first thought feel the same, they are actually very far from the same. *Why* motives can have successive generations of reasoning attached thus demonstrating high levels of intellect. *How*, on the other hand, only lets us drill down to root mechanics of any previous action.

Science, or in this case single-source long-age evolution, only ever attempts to explain the *how* aspects of creation. Where proponents of guided creation, on the other hand, typically straddle the *how/why* fence a bit but do not realize that they are doing so. Always work to differentiate the *how* and the *why* aspects when discussing the creation topic, especially in any debate forum regardless of which side you are on. A tell-tale sign of a weak argument is when you feel compelled to try to make a fool of your opponent. If they make a fool of themselves, then that is up to them. Knowing the difference between *how* and *why* can help you to not step on the opposition's land-word mines.

Common Ancestors In a Single Cell

So far, every proponent of evolution that I have spoken with has indicated to me that all living creatures came from a single source. That is to say that lightning striking the gases in the right conditions created the amino acids that led to a single cell eventually being created and that everything came from that one single cell. There is no doubt about it that if this is actually the way it happened, that every living creature would in fact be from that one single cell. But is this really a logical explanation? No, not if you really think about it. Now if an evolutionist said that this occurred all over the Earth and there was not one but maybe millions of instances of this occurrence, then it might seem more plausible. However, that would present a serious problem in the entire evolution model. This is because evolution is built on this long-age single-source model. The scientifically devastating part about having more than one instance of the phenomenon of lightning striking any chemicals to produce amino acids ultimately leading to cells forming, is that it would destroy the tree of life that has been invented. Any potential of multiple occurrences leads to considering millions of occurrences, and thus allowing for specific individual lineages that are completely unrelated to have occurred. Thus, birds could come from one strike, cattle from another strike, fish from another, etc. with no end in sight regarding the species divisions.

Science accepting this in any way discredits the vast majority of the modern progressive evolutionary model typically used. It cuts down the tree of life and plants a forest of life, which in reality is far more likely. But if that where the case where many lightning strikes around the globe caused initial cells to form ultimately resulting in various species, then we have to weigh the odds of very similar biology being created in the many randomly occurring scenarios. This might seem acceptable on the surface and could allow evolution and creation to actually come a bit closer together, until you consider the functionality and mechanics of DNA and how it works and how it replicates itself. And this is why single-cell origins are so very important to the typical evolution model. Multiple single-cell starts around the globe cause a demand for explanation of the structural consistency of cells and DNA functionality.

Two of Every Kind

When it comes to typical creation perspectives, a great deal of unanswerable creation questions end in answers of, "that's just the way God did it." In essence answering the *why*, but leaving the *how* untouched. Fair enough—If you don't know just admit it, or maybe offer a speculation and indicate it as a speculation. Most people can accept and respect that.

When it comes to evolution, the male/female dichotomy does present some problems. In such cases what usually occurs is that examples of creatures that might resemble male or female anatomy in some sort of functional way are offered as evidence. Such examples are interesting, and potentially possible, but generally unlikely. Just because something appears as similar does not imply that there is any provable connection. This is an enormous problem with the so-called "tree of life" or more currently referred to as a "cladogram". There are a good amount of creatures that have not yet found their way on to that cladogram, as if they are unwanted friends deliberately forgotten, but eventually they will find their place.

Up to this point in time, there seems to be no real good explanation of how the genders came to be from the evolution side of the debate, although there are vague and foggy suggestions that typically gloss over that issue. This is truly unfortunate because each *pair* in the "kinds" is how a species continues, and it seems almost senseless to continue down the evolution trail scientifically until we can get to the core of the male/female dichotomy.

Chapter 7

Making the Right Separations

Earlier in this book the importance of being able to make mental divisions or separations was discussed. Making distinctions is truly important in the analysis, or maybe better put, the troubleshooting, of any scientific endeavor. However, if these distinctions or separations are improperly placed in our minds it will have ramifications down the line on all subsequent thinking regarding the subject being studied. The standard evolution model is a classic example of this; we also have seen this in the astrophysics realm, where if you happen to be *properly* assessing "red-shift" then everything about the big bang model will prove to be incorrect.

As discussed in the last chapter, the primary leg on which evolution stands is the faux distinction that everything came from a single-source, rather than from multiple sources. Most charts, estimates, theories, etc. are based upon that premise. But what if that fundamental premise is wrong?

"cus" Words of Phylogeny

As previously alluded to, you will be inundated with evolution jargon when diving into the murky evolution pool. There is nothing wrong with industry-specific jargon; other than it makes it much more difficult for outside people to understand the points being made with that jargon. Most of those terms are not needed by the general public who are able to understand without knowing the specific definitions, so long as the underlying concepts being explained that the terms intend to convey are understood.

One element of this jargon is to use an ancient or dead language that is no longer used or rarely changes, for the purpose of naming parts of things or an archeological find itself. The reason for this is that we can give something a name derived from the dead language and, in theory, the meaning will stay consistent because the language will not change because it is generally no longer used. You can think of this in terms of "hey, that's cool man!" versus "wow it is really cool outside today.", where the meaning of "cool" has changed between the two statements.

With evolution we use terms like "archaeopteryx" basically meaning *ancient feather* from Greek, or the Latin/Greek combination "australopithecus" basically meaning *southern ape*, which is a better fit for *The Science Of God Volume 4 – Day Six – Evolution versus Man – In Our Image*. Adding "cus" or "us" or "is" or "sis" or other such suffixes in order to make something feel more authoritative does have the desired effect. It is obviously more than just adding such suffixes, but when combining Latin and Greek to name an archeological find, it offers evolutionists a wide variety of options with which to woo the world. So as to not be too smitten with these terms and their implied meanings, always look behind the curtain to see what they mean in your native language, because they are usually little more than location and creature type that have been Greek- or Latin-ized. No disrespect for these names is intended, but as someone who is looking into this, it is good for you to understand what it is that the name intends to convey to you in order to understand what the specimen is. These names are typically invented by the person who found the specimen and that naming is arbitrary to the decision of the person doing the naming done in our contemporary time. This means that it is *not* some authoritative name of old or that it has been approved by a discerning group of scientists, but rather it is done at the discretion of the finder. Do not be impressed by or smitten with these names.

Systematic Classification of Life

Earlier the idea of a single-source or common original ancestor was mentioned, and there are many evolutionists who blindly accept that single source evolution is logical. It is important to understand that if you are serious and want to have serious discussions about the single-source, you will be quickly defeated, even by some fellow evolutionists. The "systematic classification" system is complex and there is much DNA or RNA crosslinking implied by it in the early stages of life, making multiple starts the more likely possibility. Of course this is largely speculative, but far more convincing than single-source theories. This means that it would be impossible to detect if a single-source ever did exist because there is so much early intermixing implied in the multi-trunk theory. This is not the same as multiple single-sources occurring around the globe as mentioned earlier. In this theory of early intermixing, there very well could have been a single-source, but as the replications and deviations occurred, intermingling of the various subsequent deviations would have also have occurred very early on causing an untraceable mix of the early stages of life. Such theories, accurate or inaccurate, are created to compensate for the unlikely possibility that everything came from a single-source, and such theories are neither falsifiable nor verifiable.

This allows an exception for the need for having to explain the initial point of evolution, thus it is not verifiable or falsifiable. This suggested initial part of creature intermixing is a point of blind faith. Now, depending upon how long this book that you are now reading stands the test of time, the evolutionary theories relayed here are likely to change and possibly be discarded altogether in decades to come because the evolutionary theory models evolve far quicker than they claim the creatures do. Unless we can arrive at some undeniable proof, these evolution model changes will continue to occur as one aspect or another is found to contain compromised logic, or has been altogether disproven.

The systematic classification system is based on the fossils found in the various rock layers around the world, as well as through the radiometric dating and/or estimates of the age of the layers the items are found in. In theory, the deeper you go in the layers the simpler the fossils that are found are, meaning that they have less and less complexity the deeper you go. This does indeed seem logical; however, it is an extremely fragile position.

The fragility of this position is not fragile in general, but rather it has single critical vulnerability which is, if ever anyone was to discover a much more complex life form at an older or lower layer, it nullifies the entire evolutionary model of the currently found fossils. Such a find would ultimately devastate the entire evolution community. But we can expect that would be the point where the various layers would no longer be considered reliably deposited across the globe. Only time will tell if such a find will occur, but I suspect it will be found.

At this point in time the layers around the world can be questioned regarding both age and continental alignment when gauging items found from one continent to the next. But in general, few people are willing to challenge the very rigid belief system of the ages of the worldwide layer/periods suggested in the geological column. They might be very accurate, or it could be a great deal of guesswork regarding the age of the layers.

Defining Phylogeny

"Phylogeny" is a branch of biology studying "phylogenesis" which is a Greek derivative for *tribal-origin*. It is best described as the biological process from which a taxon appears. The subject is referred to as "phylogenetics".

Challenge Phylogeny

People are invited to challenge the phylogeny model to see if they can find any breaches in the chain/tree of succession. It is a vast tree with many branches that would take a one person a very long time to fully dismantle. This tree is not built by a single person, but is supplemented by many evolutionists. And every breach found it in will be quickly patched, so trying to challenge it is a monumental task. An important aspect of this tree is that there are very obvious connections between various creatures in that tree. Take for instance various breeds of cats; we know and have absolute evidence that cats do change when they are crossbred to create new breeds, thus proving that evolution does at least exist to some extent. These smaller evolutionary steps will for very good reason appear on any phylogenic or genetic chart, but they are mixed with suggested transitions that are not particularly provable, such as those that occurred thousands, tens of thousands, or even millions or billions of years ago.

With regard to the older specimens, it is critical that we are able to separate the speculative from those that are actually documentable by actually witnessing them occur. Even though such a suggestion will be met with hostility, it is critical to make the distinction in your own mind between *provable* and *suggested* when viewing these charts. There are many mental separations that you will discover need to be made. Some more critical separations are that of pictorial charts with artist renderings versus numbered or coded charts.

Numbered/coded charts that list names or codes for the various clades etc. are complex and require you to actually verify the items by finding pictures of the actual find and then comparing it to the items coded in the neighboring and parent branches. While the identification code or number has some relevance to the system, at first glance it appears to be arbitrary codes/names. With the vast makeup of these charts, looking up every code to verify them is a monumental task for any one person.

The pictorial charts are a bit easier to conceive but they all share a common fault, which is that they all contain artists' renderings of what artists imagined the creature looked like. There is obviously little other choice because we do not have the actual live creature here today to use as a model because that specific creature type allegedly no longer exists. The implications of this are that the visually physical features depicted are in essence nonsensical, so we must then revert to the actual fossils found and determine if they could logically fall within the chain of linear succession in which they are placed, but this offers a greater level of difficulty as researching coded charts.

These charts are to an evolutionist what the Bible is to a creation supporter. However, they do differ from the Bible in that the Bible is, without question, composed of ancient documents that are compiled into a single book that evolution supporters have been continually working to vilify. But believers have been working to verify and reinforce Bible translation integrity for thousands of years. Versus the phylogeny charts completely made in recent times that are based largely on speculation and are a mix of a small amount of provable witnessable facts and a vast amount of speculation regarding the ages and meaning of the many fossils found.

Common Ancestor

"Kind" versus "species" is often a stumbling point as mentioned in a previous chapter. If you partake in discussions about creation versus evolution there are two key aspects or separations that need to be made. The first separation is the actual Biblical account versus the scientific account, and the second is the "kind" divisions mentioned in Genesis versus the species or clades mentioned in evolution. The scope of data in Genesis is quite brief when contrasted with the hair-splitting that occurs when building the modern evolutionary tree.

Can these two perspectives co-exist? Do either of them indicate common ancestry? To an extent the answer is yes and yes. But here again we are met with a separation or definition requirement. Creation supporters often wince at the mention of evolution, and instead they will refer to it as micro-changes or micro-evolution, making the point that evolution is limited in scope. Some creation supporters will agree that a single-source ancestor might have existed but only for a certain "kind", indicating that another "kind" will have a different single-source ancestor. They will generally reject any idea that the various kinds came from only one single-source point of origin. Whereas evolution supporters have little problem accepting single-source origins, and to explain the unexplainable they allow the idea of early stage inter-clade cross mutating.

It does seem that if we can get over our defensiveness it can be admitted that the idea of a one single-source origin is somewhat unlikely and is illogical regardless of which discipline we choose to study or believe.

Chapter 8

The Bodies of Evidence

Evidence is important to any scientific endeavor. This is true whether it is strictly scientific or even if it's Biblical science. What evidence do we have that the Biblical creation account is accurate? And what evidence do we have that the strictly scientific origins account is accurate?

Biblesaurus

On more than one occasion I have heard people asking: We have evidence that large creatures such as dinosaurs existed, so why are dinosaurs not named in the Bible? That is a very fair question and deserves a full answer.

To begin to answer that we must first dissect the word "dinosaur", what does it mean? The "dino" part means *huge* or *terrifying*. And the "saur" part means *lizard*. "Dinosaur" is one of those Greek-ized type description names mentioned in a previous chapter.

Next we need to determine how old the term "Dinosaur" is. As far as I could determine, the first use of the term "dinosaur" occurred somewhere during the first half of the nineteenth century. This would coincide with the explosion of discovery around that time due to the fervor for finding fossils. When we say "dinosaur" we are doing a couple of things: The first is that we are using a relatively *new* term when compared to the age of the Bible. We know that the books in the Bible are a minimum of two thousand years old, and some much older than that. There is a great amount of historical writing on that subject along with a great deal of matching evidence found in many archeological digs that confirm many parts of the Bible. So in considering the relatively new term *dinosaur* we must understand that it is relatively new (See *Understanding The Bible - The Bible How-To Manual* AND *The Things We Don't See*).

The second aspect of the word is its *assumed* meaning of "*terrible lizard*". What we don't know about dinosaurs is what they actually looked like. **Every** picture we see of them is an artist's rendering, or what they imagined the fossil remains would have looked like while the creature was alive. These artists' drawings might be very accurate, or they could have some major flaws. We may never know for sure. The reason I bring this up is that they might not have looked like lizards at all, though we can measure the fossils so we do know logically that these creatures were very large or terrifying in size. A more accurate name might be an English representation of Greek, being *dinoplasma* meaning *terrifying creature* or *terrifying form*.

The assumption that they are lizard-like is not necessarily wrong, but it does cement things in our minds to a very specific look. But, we must realize that not all "dinosaurs" look like lizards in the artist renderings, thus using "dinosaur" in our modern context as an overall description of these large creatures borders on dishonest. You could think of this in terms of some of them being a bit more like the look of an elephant regarding their skin and massive appearance. So regarding the question of whether or not dinosaurs are mentioned in the Bible, they most certainly are!

The most prominent mention is in the book of Job where it describes the "behemoth" and the "leviathan". If the book of Job was written even as only a story, and we know that it is at least a couple of thousand years old, the fact that someone mentioned very specific attributes of the large creatures is evidence that someone back then was familiar with them in some way. But since we are mostly concerned with days five and six of creation, we need to back up a bunch of sections in the Bible all the way back to Genesis One. Does the Genesis creation account mention any large creatures? Here is that portion of the text again for your evaluation:

Douay English Genesis One:

"[20] God also said: Let the waters bring forth the creeping creature having life, and the fowl that may fly over the earth under the firmament of heaven. [21] And God created the great whales, and every living and moving creature, which the waters brought forth, according to their kinds, and every winged fowl according to its kind. And God saw that it was good. [22] And he blessed them, saying: Increase and multiply, and fill the waters of the sea: and let the birds be multiplied upon the earth. [23] And the evening and morning were the fifth day.

[24] And God said: Let the earth bring forth the living creature in its kind, cattle and creeping things, and beasts of the earth, according to their kinds. And it was so done. [25] And God made the beasts of the earth according to their kinds, and cattle, and everything that creepeth on the earth after its kind. And God saw that it was good."

In reading the text just shown, we can see that there are a couple of places that "dinosaurs" could be included in the text as it stands. "the creeping creature having life" and "every living and moving creature" are a very broad all-inclusive statements listed for day five. Then on day six it says "the living creature in its kind, cattle and creeping things, and beasts of the earth" this is another very broad statement that unquestionably allows for "dinosaurs".

However, on day five there is a bit of a translation trap that we should examine. The types of creature listed are very broad with the exception of the "whale". *Beasts, creeping things*, and *birds* are all very simple and broad categories, but *whale* is not. While there are a variety of whales in the oceans, it is still a very specific listing uncharacteristically mixed together with other very non-specific creature categories. It is unlikely that "Whale" is what was originally written or intended in Genesis One, as was mentioned in an earlier chapter.

Douay English Genesis 1:21

"And God created the **great whales**, and every living and moving creature, which the waters brought forth, according to their kinds, and every winged fowl according to its kind. And God saw that it was good."

Latin Vulgate Genesis 1:21

"Creavitque Deus **cete grandia**, et omnem animam viventem atque motabilem, quam produxerant aquae in species suas, et omne volatile secundum genus suum. Et vidit Deus quod esset bonum."

Here you will notice that the Latin does not say "whale", but it does say "grandia" clearly implying *large*. In the case of the translation to "whale", we immediately run into translation indoctrination problems. The term "whale" was not randomly chosen. When translating from Latin, "cete" does translate to "whale" or "sea creature". So in this case we need to back up the translation bus back to at least the Greek Septuagint. In Greek the word would sound something like "kaytos" which is often translated as *sea monster, whale,* or *huge fish*. It is at this point in the translation chain that we get somewhat stuck. Since we do not have the actual original Hebrew text, we can only visually review these older Greek and Latin Bible versions, respectively the Septuagint and the Jerome Vulgate. People often want to use the available Hebrew/Masoretic text, but that is actually not as old as the Greek and Latin versions are. There is a commonly passed along error that these Bibles with the Hebrew script are the best source to study, but that is not completely true, because the chances are very strong that that text is translated from the

Latin. There are many unknowns in this regard, with a lot of speculation.

So, back to the whale of a problem on day five. Since the other groupings are all quite broad, we can be reasonably logically sure that "whale" is a poor translation choice. "sea monster" is likely a much better choice of description. However, "sea monster" is a somewhat redundant implication because it is typically translated along with the word "great" implying very large. And the term *monster* typically implies either grotesque or huge. From this we can make the reasonable conclusion that it is referring to large creatures. But we are stuck with the issue of this description typically being referred to as some sort of water-dependent creature, ie a *whale* or *sea creature* or *sea monster*. When looking at the Latin "Cete" or the Greek "kaytos" both tend to indicate a water creature, and this is very possible, but what we do not know is what the original text actually indicated. We can to some extent assume that the thought of them being water creatures is due to the fact that it is the water that brought them forth, but imagine if we abandon the idea that these huge creatures are water creatures only since the birds are not, then we open the possibility that it includes all of the various types of "dinosaurs" as well.

You might wonder why so much time would be spent dissecting this particular aspect. It is because of the peculiar nature of the specificity of the term "whale" that does not fit with the creation account's broad descriptions used everywhere else. Further, as alluded to earlier, the creatures described on day five are generally egg-laying creatures, and "whales" typically are not. This makes "whales" even further off base. However, dinosaurs as far as we can tell, did lay eggs and would be a far better fit. To add to this idea, a fire breathing dragon or "Leviathan" is a large creature that is described in Job as being a water creature.

Some of what is mentioned here is speculation where other aspects are not. Regardless, using the term "giant sea creature" or simply "giant creature" is likely the most accurate way to translate this particular issue, and doing so is within the realm of "whale", "monster", and "dinosaur". But we still have to question the issue of the distinction being between egg-laying creatures versus mammals between day five and day six. We will examine that more later. If the creation account is true, then egg-laying dinosaurs were likely brought forth by the waters on day five as that is a very reasonable and comfortable fit for these creatures that we have little idea of how they actually looked, but we have compelling evidence that some did lay eggs.

Scientific History

"Science" is a bit of a complex topic to tackle. When did "science" begin, meaning how old is it? That depends on what "science" actually is. Based on the word's etymology, the word "science" means to *know* or to *split*. In our modern time we tend to feel that science began during the so-called enlightenment period, but I question that thought. The first known use of the word was in the fourteenth century. But while prior to that time it might not have been referred to as "science", the idea we call "science" certainly did exist and it was generally those who were dedicated to God, who sought to know and understand the God who did what we today call "science".

There is wrong thinking that most people have regarding people from the old ancient days where we assume that they did not know things. As mentioned in previous versions of *The Science Of God*, too often we have this belief that things have only been known for the past few hundred years, yet all evidence shows us otherwise. Consider "flat Earth theory"; for instance, we believe that people believed that the Earth was flat, yet ancient artifacts and recordings indicate otherwise. In fact as the story goes, Christopher Columbus realized this and thus was not afraid to set sail for distant lands. He did not fear falling off the edge of the Earth.

It is only when someone has an erred theory that appears plausible to any contemporary civilization that we deceive ourselves. If someone makes an argument that the world is flat and it seems indisputable then it is likely that many people will fall for the errors and believe it to be true and will subsequently promote it as true—but that does not make it true. Then later we look at actual evidence rather than following someone else's absurd theory and then we find that the earth is not flat and has no edges to fall off of and then we believe that *we* figured it out and that all previous civilizations did not know what we now know. This is ridiculous and arrogant. Evidence of long past civilizations indicates that the world was round at minimum and we can even deduce that many knew that it was spherical, but in our modern arrogance we believe that *we* have only just figured it out. We do not want to believe that we believed wrongly about this in our relatively modern era.

Darwinian evolution is one of those things. Darwinian evolution is a convincing argument that has lured many people into its errors. Does this mean evolution did not occur? No, but it also does not mean that it did happen, especially single-source long-age evolution. There is a big difference between something theoretically being able to occur and something actually occurring. The error point of long-age Darwinian evolution is that because predictions are believed to have been proven "true" it gives the illusion that evolution has occurred, but this overlooks a great deal of other possibilities.

Science is not new, and in the Renaissance period the *church* funded a great deal of science. People of ancient days obviously understood science and math and were quite proficient at it all. There are many ancient ruins and remaining ancient structures that, to this day, we still cannot figure out how they were built. The technique of science is very old, and most of what we know and our passion for it has to do with the Biblical "western" culture and zeal for trying to understand God the Creator, as is clearly made evident when studying recent and ancient history.

The Christian and How It All Came to Be

It is a ridiculous approach to insist that just because some people view Biblical creation one way that all people who support the creation account understand it in the same way. It is the same with evolution supporters; make no mistake about it, not all evolutionists believe or understand evolution through the pop-science evolutionary lens.

Generally, those evolutionists who are hostile (not all are, but it is common) will group *all* Christians in a six-twenty-four-hour-day creation group. And further, as experienced firsthand, they will insist that you believe the Bible says what they say it says. But this is a very agenda-driven goal, because if the Bible doesn't say what they have chosen to believe it says, then it would have major ramifications for their entire blind faith pop-science belief system. Often this type of evolutionist has blind faith regarding the Bible equal to that of the six-twenty-four-hour-day creationist, with the caveat that they in their blind faith specifically do ***not*** believe the Bible, and they don't believe it in an improper manner, which was discussed earlier where they were typically taught six-twenty-four-hour-day creation as children and believe that is what the Bible indicates, even though it does not.

If you are a creation supporter or an evolution supporter, be cautious in your interpretation of other people's understanding of the Genesis creation text, because not every person understands the text in the same way you do. And further, realize that it is not only Christians that support the guided creation models, Jews and Muslims also believe this, and it is not a small number of people. Further, there are many cultures that have similar creation stories, but with some unlikely deviations. The point here is that the godless-Darwinian-evolution model is a *minority* view in our world. And areas such as communist China must force their people to *not* believe the Bible, which is very telling of the broader evolutionary agenda.

Dinosaur Pictures and Other Figures

So as to not let this one slip by, we will briefly revisit our mental images of dinosaurs. Just as few people actually read their holy books completely cover-to-cover, so too do people not read the scientific data completely. The Quran is a very short book but very few Muslims have read it cover to cover, especially the radical Muslims, or they would obviously not be so radical. Similarly, there are very few Christians that have read the entire Bible if they have read any of it at all, and it is little different with Jews. We people are by and large quite lazy when it comes to studying. We just want someone to tell us what it is all about, or make a movie about it so we can watch it in an hour and become "armchair experts".

In the case of dinosaurs, we want to look at the creative pictures drawn by creative artists. Many of these pictures could be accurate *or* they could be completely off base, we may never know for sure. But one interesting aspect of dinosaur pictures is that they are nothing new. There are allegedly cave drawings that are believed to be quite old that depict dinosaur-like creatures, though typically not nearly as detailed as we draw these days. This is a clear indication of one of two things: First, that the cave-artists may have been archeologists and dug up dinosaur carcasses, or second, they possibly walked along with dinosaurs and therefore witnessed them firsthand and relayed that to us with their drawings. Or we could imagine a third choice that they simply invented the pictures with zero mental outside influence, but that's not very likely since they match so well to our actual modern-day findings.

Because our modern-era "science" occurring since the eighteen-hundreds has so deeply embedded the age estimates of the fossil remains of dinosaurs in us, we completely discount the idea that man ever could have existed alongside dinosaurs. But let us at least consider a few things that contradict that: First we have the Bible's very large and fierce creature description in the book of Job. Next we have the tales of brave medieval knights slaying dragons. Then we have various cave drawings depicting unique creatures that appear very similar in general design as we today depict various dinosaur types. And finally, there are ancient statues and what appear to be ancient clay children's toys, and if I recall properly even brass or bronze statutes of similar looking creatures among other things found. This is a very strong argument that humans with strong intellect did live during the lifetime of at least some dinosaur types. We can also add to this that the Genesis One text generically describes large creatures of some sort.

Chapter 9

The Big and Small

Creatures big and small have walked this Earth as is made evident by the vast array of fossils that have been discovered over recent centuries. One of the questions we have to ask is when and why did these very large creatures disappear from the face of the Earth? This is the point where all of the various historical geological periods are introduced. As of the writing of this book the pop-science consensus is that an asteroid struck the Earth and caused catastrophic devastation that completely wiped out all of the dinosaurs and larger extinct animals, though there are some variances in those theories. This convenient non-falsifiable theory explains a lot, or maybe better stated it covers recent dinosaur tracks.

It is obvious that meteors and asteroids impacted Earth in the past, but we cannot *prove* that any of them caused the extinction of dinosaurs. How did the large creatures become extinct through this catastrophic event but not the smaller creatures, where it targeted only the fairly large ones that are dinosaur-like? How did creatures such as the hippopotamus or elephant or the mammoth and others survive long after? The asteroid or meteor theory doesn't hold up logically. Some people claim that the black layer proves a massive asteroid hit that killed the dinosaurs, but that defies logic because then the black layers would have mass extinction in subsequent adjacent layers and within itself, and this is not where most dinosaur remnants are found.

Scaling Things Down

Why are there no other super-large creatures remaining alive beyond those in the ocean today? Did they become extinct or did they evolve into smaller creatures? There's a lot that we do not know or understand about "prehistoric" life. We simply do not know if things changed such as the rotation speed of the Earth, or atmospheric pressure, or the rotation angle of the Earth, or the duration of a year in days, or Earth's distance from the Sun. Some scientific evidence indicates that some of these things could have substantially changed over the vast history of Earth. One of the more telling experiments I read about was a test done raising fish in an atmospheric pressure that is far different than what we experience today. In this experiment the fish grew much larger in size than the same type fish do under our current atmospheric pressure conditions.

The implications of creatures being able to grow larger than is normal to us under an altered atmospheric pressure are substantial. If this experiment is reliably repeatable and if we could find any reason that Earth's atmospheric pressure has substantially changed over time, it could explain how there are so many enormous fossils found that are similar to, but much larger than, creatures that exist today.

Pollywogs

"Evolution" is change over time from generation to generation in the lineage of any creature. When there is a gap in the evolution charts, people naturally want to fill that gap in. If we need to morph from a large fish, such as a whale to a crocodile, then we need to find interim fossils that would meet those interim form requirements. Often after a great deal of seeking, these fossils are rarely found. Presenting "Pakicetus"! (Search for images of Pakicetus to see if you can imagine it as a transitional fossil.) However, those findings do sometimes stretch our imaginations quite a bit in order to fill the empty phylogeny slot where eager evolutionists are all too happy to welcome the sometimes-questionable creature's form into that slot.

One point of evolution that was sought was a transitional fossil that would demonstrate the morphing required to bridge the gap between a legged creature to or from a whale or large fish, and this questionable fossil was indeed found as just mentioned, and was named "Pakicetus". Yet, an immediate morph of a pollywog which resembles a fish will within its own lifetime grow legs and morph into a frog. Could this also have occurred with large creatures in their own lifespan as they grew while in their infantile stages of life? If so, we could find such an ancient creature in various stages of its morphing form and completely misinterpret the fossil data by assuming it evolved when those creatures only morphed in a similar way as a pollywog does within its lifetime. Yet even with that and the actual fossil Pakicetus found, the lineage connection is very vague at best. To put it bluntly, Pakicetus being a transitional fossil from whale to for legged creature is a connection that is stretched far beyond a reasonable imagination.

Location, Location, Location

Assessing the landscape can be a difficult task when trying to find dinosaur fossils. One of the cleanest finds was a T-Rex found around the "Badlands" near Faith, South Dakota at an elevation of roughly twenty-six-hundred feet above sea level. There are areas where people tend to find dinosaurs more readily. Elevation of a fossil is a partial indicator used to estimate the age of the deposited layer(s) in which the fossil was found. In other words, where there is one you might find others if allowed to dig more. Unfortunately, in the case of the South Dakota T-Rex, the government confiscated it from the finders due to "land rights" issues. A very suspicious act by the government, almost as if they were trying to hide something. If you are interested in knowing more about this T-Rex, look up "Sue the t-rex".

When looking for fossils we must consider what we are looking for and what that creature might have needed to survive and what its migration habits might have been. We must also try to find a geographic location that would match those needs, and finally we must start digging, a lot.

Bird, Brains, and Bones

Dinosaurs are believed to be the ancestors of birds because they share certain design attributes, this reinforces the day five categorization point made in a previous chapter regarding dinosaurs being included in the day five creation events. It is very logical that the creatures created on a given day would have some shared characteristics that are unique.

Discounting the nonsensical connect between Archaeopteryx and dinosaurs, some dinosaurs appear to share the hollow bone aspects much the way birds exhibit. But not all dinosaurs have hollow bones, so this is not a rule and it does not prove design connection. The issue of archaeopteryx the dino-bird partly defies direct lineage from dinosaurs to birds or vice versa because they are believed to have existed around the same time, thus one could not have come from the other, but rather potentially would have a shared ancestor, or in the case of creation, they are possibly of the same DNA design set that was established on day five.

Chapter 10

Evolution of Species

As discussed earlier, species designation can be a complex topic with many branches. Because the definitions for *species* are quite varied and broad, new terms have been inserted into the evolution models to better categorize branches of creature deviation.

Families and Kingdoms Don't Exist?

Origins scientists, long since passed, used terms like *families* and *kingdoms* and *species* etc. However, those terms have more of a Biblical connotation and have generally been rejected in more recent times and are replaced with words like *clades* etc. Biblically speaking, you could name the broad groups and call them *families* or *kingdoms*, but you still only have those very few that are listed in Genesis One on days five and six. And since there are so few categories mentioned in Genesis, labeling them as *families* or *kingdoms* is not needed. In Genesis we have *birds, creeping creatures, giant creatures, cattle, and every living and moving creature*. Days five and six truly do cover it all in a very broad sense. The modern divisions that we are placing on the Biblical "kinds" are man's excessive divisions, and such detailed divisions are certainly *not* discussed in Genesis One.

There is one thing that is commanded on days five and six regarding all of these creatures and that is that they will all be brought forth "after their kind", this is inarguable fact. This means that a cat is not going to spontaneously give birth to an elephant or a frog, it will give birth to a cat. This model holds true everywhere we look in everything we see and every fossil we have found. This of course says nothing of possible gradual changes over time. Thus "after its kind" does not specifically preclude the general theory of evolution.

Finding New Evidence

We frequently find new evidence regardless of how we interpret that information. Often when an artifact is found, it might be a tooth or a femur bone or a tusk, much data can be extracted from that find—maybe too much. Speculating when finding fossils is fine, but are there reasonable limits as to the meaning of a found piece?

There was once a documentary made regarding a tooth that was found somewhere in Alaska. Keep in mind that this was a *single* tooth. The documentary went into great detail about the kind of animal it belonged to and even presented pictures of this animal. They explained how it migrated. They also discussed its diet and much more. How they extracted all of this information from a single tooth is quite a mystery. These sorts of "documentary" programs are presented to unsuspecting impressionable school children who expect to get truth from the programs we adults make *and* allow them to watch.

This was perhaps the most dishonest "documentary" touching evolution ever witnessed. A great deal of information can be obtained from a tooth, generally the type of animal it is from, and from that we can speculate how such animals typically lived, but to create an animated model that deviates from what we know today based only on a tooth, is a lie. And people who present this garbage should not have their theories forwarded. The unfortunate problem that we deal with in these days of 24-hour television programming is that these broadcast, cable, and internet networks will do anything for a buck. And since the programming is often broadcast twenty-four hours a day, seven days a week, they are desperate to create new content, so any lie will do as long as it fills a time slot and pays the bills. Shield your own and other children from such lies, or better yet, reveal to them the outright dishonesty in such programming and textbooks. Let's stick to the facts and call speculation what is–speculation.

Definition of Species

Here, let us take a moment to define "species". Not specifically the various species, but rather the implications of a species. Since the various science disciplines have a wide variety of species definitions, by just grabbing a definition we immediately run into definition conflicts. So here we will take a brief look at word origin to see what the word "species" was originally trying to convey to us. As best as can be understood, the etymology in Webster's dictionary is *"kind, appearance, species, or specere meaning to look or spy"*. It basically means to look at something and find discernable differences. If the differences are strong enough, then it can be considered a different "species". So in a loose sense, crossbreeding two very different breeds of cats creates a uniquely different cat that can then be considered a "species", thus proving that evolution is relatively fast-acting. There should at least be some sort of consistency in defining *clades* and especially *species*, but at some point a determination must be made as to whether or not the deviation is worthy of recording as a unique entity type.

The Idea of the ancestral divisions of "clades" is a never-ending tree that includes *every* creature that every existed. Much like a family tree, you would record every single generation of every single creature for it to be complete because we do not know how much a creature will deviate in say twenty or thirty generations.

Including every creature would be impossible, thus the phylogenic tree will never be complete, which is obvious in that case. "Species" is different because it is more specific in that it attempts to classify broader groups more rigidly. But who gets to make those classification decisions? We could say that each species is a clade, but there might not be enough of a difference from one clade to the next to differentiate them as someone who is categorizing a species would like there to be.

Stop Crossing Species

The creation-versus-evolution topic is a difficult topic because it is wrought with emotion stemming from a belief in God or the rejection thereof. This division causes a lot of heated arguments and causes people to close their eyes and ears to good new information from their opposition.

The scientific inconsistencies in defining species by many evolutionists are somewhat frustrating to creation supporters because such moving targets are difficult to debate. Evolutionists do seem to like it that way, but is it any better when creation supporters try to insist on kinds or species that step beyond the scope of the Genesis One creation account? It is not.

Regarding the human species versus the ape species, we break down apes in to groups that are less defined than the various human ethnicities; In this case we are essentially crossing species. That's not the only inconsistency in this particular evolution area, but it's an obvious one. Classifying species in an inconsistent manner, regardless of which side you debate from, is annoying and frustrating. And from a Biblical creation standpoint, it is foolish since the Bible's Genesis One is quite limited in that regard.

Blatantly Dishonest

When it comes to these debates and the various arguments offered by each side, there are some really solid views. However, many of the remaining views are simply either very ignorant or incredibly dishonest. It could be that it is all just being naive or being ignorant, but on the science side of the debate it is common that there is opportunity for tremendous financial gain that regardless of what nonsense you spew, you will be paid. This sort of financial incentive is far less on the creation side because the creation side is typically done more out of a dedication to God. Regardless of how wrong someone's theory might be, creation supporters are generally innocently naive in most cases.

By just listening to some of the obvious errors being promoted on the evolution side of the debate, it is difficult to believe these less credible views are innocent. Some theories are so far off from anything that could be considered logical that it is hard to imagine that they are not aware of the deceptions that they are promoting.

Chapter 11

Receiving the DNA

Perhaps the most important aspect of creation and/or evolution that we need to understand is DNA. Great strides forward have been made in the twentieth and twenty-first centuries regarding DNA, but as of the writing of this book we still know very little about it as will be made very clear in the upcoming decades.

We need to keep our minds open so that we do not return to our typical human antics of having a one-track mind by forcing every contradictory opinion or thought out of our minds. We need to receive the DNA information as it comes to us and adjust our thinking to match truth, rather than trying to force our thoughts into truth, which doing so never works out well in the long-run for those who do so.

Walking Fish

When fossils are found, those fossils are closely examined. And then in effort to classify them into the proper species, or clade grouping branch, they are compared with other fossils that have been found and previously classified. In addition to this the physical structure of modern creatures are also considered in the analysis of the newly found fossil. When a fossil has an unusual form or bone appendage, then we work to fit the creature into the phylogenic tree. It should be noted that possibilities of appendage birth defect anomalies are generally not entertained in evolution fossil finds, meaning that a deformity could be found and imagined to be a transition, rather than a birth anomaly.

Since we have not experienced the creature while it was alive we have little choice but to make assumptions of how it fits into our current scientific view. Regarding the fossil that this section is alluding to, we are talking about Pakicetus, which is imagined to be a "walking fish". In such determinations we must understand what is meant by "walking" and what is meant by "fish". Specifically meaning, how broad are these categories of form and function? Much the way an ostrich has wings but doesn't actually fly, so too a creature could have legs but not actually walk, and might instead slither like a snake.

Now, regarding the "fish" form, it is also a point of definition. What constitutes a "fish"? Is a fish something that must be in water, or could it be amphibious. Is a whale a fish? Is a crocodile or an alligator a fish? If you take a look at the "Tiktaalik" "walking fish" fossil you will need to examine the images of the full extents of the actual structure that is discernable from the fossil. Then compare that against the artists "restoration" rendering and decide if you think that there is enough information in that particular fossil to merit the particular "restoration" renderings you find in the world of pop-science.

Was this particular "walking fish" an evolutionary transition? Or is it a creature similar to some lizard-like creature, but a bit larger? There are known "walking fish" that exist today and these creatures may well be transitional living forms, or possibly not. But some could be linearly related to the Tiktaalik fossil. Large size fossil forms have been found that have striking resemblance to much smaller creatures that live today. Tiktaalik is likely no different. The Tiktaalik skull is considered a flat skull, but since bones can flex somewhat, especially over time and when buried, we simply do not know for sure if the skull was deformed and flattened as the weight of the subsequent deposited layers settled on top of it.

There are many assumptions being made about the Pakicetus and Tiktaalik fossils that are stretched probably a bit further than they should be stretched. The likelihood that modern living and breathing "walking fish" have received their DNA from Tiktaalik and its counterparts of the era is high. However, this does not bar Tiktaalik from being a transitional form, because evolutionarily speaking, a form does not have to phase itself out, it only needs to give rise to additional lineages or clades or species of creature. We will be discussing this later regarding the limits of creation and of evolution.

Genetic Language

Genes are assembled at conception and use a language of instructions much like words or computer code. The "computer code" or *program* is the particular arrangement of the DNA as far as we understand at this point, and it is what makes creatures take the form that they have. However, there is a great deal that we do not understand about DNA's C, G, A, and T sequence of code. As far as is understood at this time, the DNA is largely the same throughout any one creature, yet one cell will become a part of a limb and another cell will become a part of a brain or some other organ. We are not yet certain why a leg becomes a leg, or an arm becomes an arm, thought it does appear that we are getting closer to understanding some of these functions of DNA, yet there is still a long road ahead before we grasp it all, if we ever do.

Lost Information

Many creation supporters claim that DNA information can only be lost, but with this I disagree. Information can obviously also be created and it can be changed or exchanged. Much of the creation debate is really regarding guided creation and whether or not humans evolved from some sort of primate which is what *The Science Of God Volume 4 - Day Six - Evolution versus Man - In Our Image* is about. Regardless of the creature type, a common creation belief is that a "kind" was created at the point of that creature's type/kind creation, and all of its genetic code was created at that time, Biblically speaking. The claim is that DNA or other instruction can only be *lost* from that point on. The losses of various aspects of instruction, whether DNA or otherwise, is then the cause of deviation of the specific kind thus explaining the varieties we experience today. It is possible that this is true, but it is unlikely because it would make the kind and its ultimate survival extremely fragile and short lived. Eventually the creatures would lose critical features like legs or eyes and eventually reproductive organs, bringing them to a speedy demise.

It seems logical that creation supporters are correct in regard to information being able to be lost. Even evolution supporters agree with that regarding DNA. What the typical creation perspective is missing is that it is very obvious that DNA is alterable and can be added to. For instance, characteristics can be bred out of an animal and then reintroduced at a later time by crossbreeding the lacking creature with a creature that still has the particular characteristic.

What We Do and Don't Know About DNA

What do we and don't we know about DNA? A lot! We can assume that as a culture we scientifically know everything about DNA, but nothing could be further from the truth. We may know a great deal, but until we are able to cure disease through

DNA or regrow teeth or limbs reliably made to order for the patient, we must admit that we know very little in comparison to what there all is to know. In our computerized world we run all instructions through our I/O or *on* and *off* sequencing and do all computing with that technique, but DNA is constructed of four options as far as we can tell, making it far more complex. At this point in time, we believe that DNA is the instruction set that determines what a creature will look like and thus determines its classification. We are also quite certain that if DNA is disrupted it can cause anomalies in the configuration of a creature by producing birth defects. We have good evidence that outside influences can negatively affect the DNA and therefore the form of the creature. But we also can assume that the reverse is true, and that good nutrition positively affects DNA to make the creature lineage stronger and healthier.

What we don't yet know is if DNA is the only way cells are being instructed to multiply to become the body part that they become a part of, or if some other force is somehow assisting and directing the cells. We also do not know for sure if DNA is self-correcting or if it will remain permanently compromised throughout *all* subsequent generations if it is in some way damaged, though it does appear that it is self-correcting to some extent. These points only touch on what we do and do not know about DNA. There is a great deal of research and speculation yet ahead and we need great minds to continue researching these things to move beyond our current sketchy understanding.

Chapter 12

What Guides the Changes

Science in general is a moving target, as it should be. Many of the talking-heads of pop-science will claim that science is always changing as it discovers and researches, but then go on to insist that very debatable things are "undeniable facts". In reality, much and even most of science is not fully decided and is always under consideration. We must speculate and then seek evidence of our speculation in order to support it. Those speculations often interfere with our conclusions and reality itself.

Moving Theories

Earlier we discussed how difficult it is to hit a moving target. To the frustration of creation supporters, evolution theory changes over time as it must in order to correct its errors if it is to be truly scientific. The frustration comes in when we generally know this, but the evolution promoters come on the scene and insist that it is all figured out and that everything is based on "undeniable fact" when that simply is not true. Fact and truth do not change, but our findings and conclusions do. These moving targets become hard to debate because when you corner them with logic to point out an error, you will often be directed to a lengthy book that does nothing to further their point. But if that person is honest they will eventually adjust their theory to compensate for the newly revealed anomalies. In that case the creation person will have to reevaluate the newly updated evolution theory that they then may or may not agree with. This has been an ongoing issue in the creation-versus-evolution debate and we all need to realize that the moving target is a requirement if we are honestly seeking truth.

It's somewhat different on the creation side, because to the creation person the Bible is the standard and it cannot change as long as the text is not compromised. So in the case of creation, what happens is that there generally are a small handful of prominent theories that typically do not change. Each theory, right or wrong, for the most part will stand firm in resolve and each is quite easily attacked in that regard. However, because there are several creation interpretations, evolution supporters must debate against several creation fronts.

Many of the creation interpretations are based upon the person's upbringing and what they were taught throughout life. These interpretations range from six-twenty-four-hour day creation all the way to God guided long-age big bang evolution creation. This can make the creation theories appear to change, but usually what is occurring is that the theories stay constant but the person being challenged is different, making it appear that the interpretation is changing. Some creation supporters will change their interpretation if they follow closely along with the evolution theories. In such cases they adjust their interpretation of Genesis to try to match the Bible with evolution. These are the sort of problems that plague and guide the changes in those origins subjects.

Evolved from a Rock?

I have heard debaters claim that evolutionists say creatures evolved from rocks. Is this possible? To an extent it could be true due to chemistry. Not that something specifically came from the rock, but because rocks are made of minerals which are chemicals, those chemicals could ultimately be instrumental in the chemical aspects of the creation or evolution of early life. After all, the waters brought forth and the Earth brought forth. And both water and earth are filled with chemicals. As for life evolving directly from a rock, that seems to be somewhat illogical.

Logical Contradictions

We must be cautious regardless of which side of this discussion we chose to entertain. Logical contradictions abound in most theories on both sides. If a creator does indeed exist then that Creator is perhaps the perfect model of "logical" when considering the consistency of the macro and micro realms that we can observe and witness today throughout all of creation.

What is a logical contradiction? A logical contradiction is when we make a logical claim in one place to justify our theory, but then we discard that same logic in another area in order to justify the second part of the theory. You will notice this occurring on both ends of the creation-versus-evolution tug-of-war rope.

When you are reasoning through data, discipline yourself to not allow such contradiction into your thinking. If someone points out a contradiction in your rationale, then you *must* be willing to receive and analyze the point they are bringing to your attention. If you are unwilling to reevaluate your position, then *you* are the problem. This does not mean that you must change your position, but rather that you must prove that you are not using contradictory positions to further your theory.

Evolving Parts

If you are paying attention you will hear pop-scientists promoting evolution while explaining that creature types slowly evolved limbs and other parts, and that those parts can evolve out, but cannot evolve back in. This is contradictory logic. If the limbs evolved into the creatures' makeup once to become fully formed, then there is no physical or biological reason that this could not occur repeatedly should the long-term environmental circumstances require it. If we are going to accept Darwinian evolution at all, then we must consider that the changes are often guided by environment.

Evolution of appendages is obviously apparent as Darwin so articulately pointed out with is examples of bird beak lengths and various beetles. And to an extent we can see this in the crossbreeding of animals. However, crossbreeding animals is considerably different than a hummingbird lineage developing a longer beak or tongue, over time, with which to extract nectar from flowers that have a deeper throat or receptacle. I am not aware of any one having patiently gone through many generations of a particular point in a clade or species and having that specific variation of creature's lineage morph to a verifiable deviation that could be considered strong enough of a change to merit its own species classification. However, it does seem to be a logical possibility that a particular type of creature could deviate in this way. But in this particular case the deviation of the attribute of the creature's lineage still falls within a broader kind or species grouping, meaning that no Biblical "kind" barriers have been breached. In other words, it is still identified as a "hummingbird".

Chapter 13

Life Forms Are Fluid

Many creation supporters are very rigid in their analysis of any creatures' ability to change over time. This rigid stance is often due to them not really taking a deep dive into thinking it all through, and it is also from what we experience with currently living creatures. Then of course on the other side of the debate we have the evolution supporters who have no limits on what a creature's lineage could morph into given enough generations.

Life-forms quite apparently tend to be very fluid in the ways they adapt to their environment. This is experienced in certain animals within their own lifetime with simple things like the thickening of a coat of fur in cooler seasons or regions, but how fluid are the various species in the bigger long-term picture?

The Oceans Brought Forth

Is it "let the <u>oceans</u> bring forth", or "let the <u>seas</u> bring forth", or "let the <u>waters</u> bring forth"? Here we get into that mess we call "translations" again. Many of the newer Bible versions state this in terms of "let the waters team with", or "swarm with" This of course is one of those perversions of the Bible's text that is mentioned in an earlier chapter of this book. The King James version and The Douay version match closely regarding this particular issue. There is an enormous difference between "teaming with" or "swarming with" versus to "bring forth". This is why so many have fallen away from the Bible. "teaming with" sounds all poetic, but does nothing for scientific explanation. However, "Let the waters bring forth" has very interesting implications that change the entire dialogue.

Douay English Genesis One text reads:

"[20] God also said: Let the waters bring forth the creeping creature having life, and the fowl that may fly over the earth under the firmament of heaven.

[21] And God created the great whales, and every living and moving creature, which the waters brought forth, according to their kinds, and every winged fowl according to its kind. And God saw that it was good. [22] And he blessed them, saying: Increase and multiply, and fill the waters of the sea: and let the birds be multiplied upon the earth. [23] And the evening and morning were the fifth day. [24]"

To "bring forth" offers us interesting scientific flexibility when speaking from a scientific creation position. The "teaming with" type verbiage places immediate limits on deviation and does not speak of creation but rather sudden hocus-pocus appearance. The older authoritative versions are more scientifically accurate, and studying those, rather than the recent centuries post-Reformation protestant Bibles, will be of great value to anyone who is interested in the origins topic.

To "bring forth" suggests a type of birthing or origin that is not well defined other than it is stating that something is coming from or being "brought forth" from the "waters". Thus, we have a tremendous amount of scientific scope here to allow for a progressive creation process.

Descent from Primitive Life

Scientifically we have an understanding that life began as a sort of primitive microscopic creature which eventually formed genes that each subsequent generation passed on down through its lineage. With each subsequent generation slight changes where introduced affecting the subsequent offspring of the generation. This suggested chain of changes through the generations is "evolution". It is stated by evolutionists that this is how all living creatures today have come to be.

To an extent this same model is true of Biblical creation with the exception that the creature forms of the initial "kinds" or "species" were fully developed from the beginning and did not have to morph over millions of years to be considered a "bird" or "creeping creature". While they might not have been immediately formed as fully-grown adult creatures, the first generation and initial point of creation for the "kind" occurred from the very start. In the typical creation model viewpoint, they did not come from an amoeba.

Regardless of which side of this subject you are on, the truth is that everything descended from primitive life. The question is not whether or not any particular creature descended. That is obvious. The question is, what specifically was the initial form of that so-called "primitive life."

Unused Appendages.

Unused appendages are an interesting issue. Do unused appendages evolve out over time? The evolution charts are arranged in such a way as to indicate just that, although there could be mistakes in the classification of some of the fossils found. The notion of the unused parts evolving out is related to the idea of atrophy, where if you do not use your muscles they tend to diminish and you lose strength. This can occur to an alarming level, which is a use-it-or-lose-it rationale. This is a logical perspective, but we must ask if there are limits to it. If muscles go unused for several generations, will the offspring of each generation have less and less muscles in that particular body part area? This would be difficult to prove on humans due to the time required for each generation to mature and reproduce. However, this could be tested with mice or insects that tend to mature very quickly. Yet we have not heard of any such work, but it would be very interesting to see attempted.

It's a bit of a stretch of the evolution imagination that an unused appendage would evolve out because it was unused. Especially since if it is unused, then how would the appendage ever have evolved in? However, there could be other more logical reasons that an appendage could eventually disappear from the lineage of a creature, including spontaneous deviation where the DNA was somehow affected and passed down to the offspring with that aspect of the DNA missing. Although that seems to be an unlikely event, the possibility does exist as we see in some humans that have DNA that spontaneously adopted a peculiar deviation. Yet, DNA seems to be self-correcting in most cases, which is evident with people who are born with no or stunted limbs who typically go on to have offspring that do **not** have that same attribute, so DNA appears to have some built-in self-correction capabilities.

Bears and Dogs Connected

Some evolution researchers claim that dogs and bears are related. This can be seen on some of the evolution charts, where creatures such as bears and dogs can appear connected via their ancestral lineage. Is it possible that these two very different creatures could have come from the same clade or species? Using an evolutionary judgement scale, both of these two particular branches could absolutely have come from a single clade. But how does that work when considering the various creation models?

The creation models do vary and depending upon which you follow, such a connection is *not* allowed. But in the broad accounts in the authoritative Bible versions detailing the day five and day six events, it is not out of the question. The two creature types do share some critical attributes. I have not studied their complete bone structures to see if they share all skeletal parts, but both do have fur and paws and their general structure is similar, though their size is quite different. From an evolution perspective this is completely acceptable, where on the other hand, from a creation perspective we have a few additional questions to answer.

These creatures would obviously have been brought forth by the Earth on day six as that is where they perfectly fit in the creation model text, so that is not in question.

Douay English Genesis One:

"[24] And God said: Let the earth bring forth the living creature in its kind, cattle and creeping things, and beasts of the earth, according to their kinds. And it was so done. [25] And God made the beasts of the earth according to their kinds, and cattle, and every thing that creepeth on the earth after its kind. And God saw that it was good."

Both dogs and bears fit very well in the "beasts of the earth, according to their kinds" category with no stretching of any part of the data given. But we then have to answer the question as to whether or not they have a shared ancestral lineage, or if they were independent lines of creature from the start. This is something that simply is not elaborated on in the Genesis One creation text. However, "according to their kinds" does allow for either a single shared lineage, or individual lineage based upon a strict reading of the broad nature of the text. This is where the evolution tree charts can be of assistance. If we can see a *legitimate* connection and can then research and examine the actual fossils that the connections are derived from then we could conceivably come to a reasonably clear understanding that the two are physically related in their birth lineage.

Here we must be cautious not to fall into the artists' renderings trap, and we must rely on the actual fossils alone to make such a determination. Further, we must also *disregard* the assumed arrangement and form of the *missing parts* of all fossils that reside in the lineage chain that would connect the bear and the dog. It is a murky trail but it is worth looking into these sorts of connections.

The great part about this whole subject is that much of the work has been done and cataloged for us, so now someone questioning such things can much more readily look into such a claim in a matter of days or weeks or months and get a reasonably clear answer. But in doing so we must dispose of our biases and look at the fossils for what they are and not what we want to them to be or what they are claimed to be. In researching such a claim, you can expect to have to disregard many of the fossils because of extreme fragmentation of those that are claimed to be in the lineage, meaning that too many parts of a creature's structure are missing in many fossils to legitimately use them in any serious detailed lineage study.

Chapter 14

Life is Intended to Be Robust

In listening to astrophysicists and evolution and creation supporters you will often hear them referring to the "delicate balance of the cosmos" and of creatures, additionally you might hear how if the circumstances varied even slightly that nothing in existence today could exist. The "Goldilocks zone" for our planet is a common astrophysics and evolution term that places the Earth in a position that is "just right" to keep a balanced distance from the Sun that will allow the support of life. This makes it appear that it is very delicate and that even a slight disturbance could upset the balance and then life as we know it would come to an abrupt end.

I completely and absolutely disagree with this absurd notion. In fact, it is quite the opposite situation. All evidence supports that both the cosmos and life itself are robust and are intended to be so. There is nothing that you can do to stop life as we know it or affect the cosmos in any substantial way that would cause it to dissolve. Often those who promote this absurd notion will look at quantum physics and say if this or that was slightly one way or another that everything would dissolve, and so to them that indicates a "delicate balance". While it may be true that if the quantum level items differed even slightly from what they are now that things would fall apart and dissolve, the fact of the matter is that they cannot be altered as hard as we might try. They are robust and are what everything is built upon making everything that they form very stable and robust—they are firmly set!

This is especially true with creature life-forms. Life wants to live, and though we may try, we cannot stop it. Somewhere there will always be a seed of life that even if we were to annihilate creatures and the human race it would not be long that creature life would spring up everywhere! Life wants to live and there is nothing about it that is delicate in the bigger picture. So while we can snuff out the life of one single instance of an animal species, we cannot snuff out life in general. Just because we think a creature is extinct does not mean that it will not or cannot make a return when circumstances allow.

White Bear

From an evolution or adaption perspective we have the polar bear or white bear versus the typical brown or black bear. Did this creature evolve, or did it robustly adapt for survival in some way to cause it to turn color? Do we know for sure it occurred at all, and further, do we know if it was from white to brown or from brown to white?

The consensus on this seems to be that the brown or black bear lineage diverted as the animal migrated to a cooler snowy climate. Then over time it more easily survived predators because the lighter its fur was the better it could hide in the snow. In this way the darker colored bears would have been much easier targets for any potential predators, and the whiter or lighter in color they became the more readily they could have survived as they blended better with the snow. This is a fundamental survival-of-the-fittest evolution model and is similar to the adaption model. But there is one little problem with this and it is that there really are very few predators of bears. While predators of bears do exist, they are few, and it is the bears that tend to be the predators in most cases. This is especially true in snowy climates. Further, the polar bears' predators in the same climate are not always white, so if they have any predators it would largely invalidate that theory.

Adapting to the Circumstances

"*Survival of the fittest*" is close to "*adapting to circumstances*", but *survival of the fittest* tends to look more towards predatory survival and survival in harsh environment. Adapting to circumstance is more in line with a giraffe that is thought by evolutionists to have obtained a long neck due to a lineage that was taller than its counterparts, thus allowing it to graze on the foliage of the tall trees. This would not necessarily mean that the lineage that it was derived from ceased to exist, but rather that it could thrive due to its taller stature, and the taller it became the better it could survive until it reached a height that would exceed the sustainable height which would then moderate its lineage's progressive changes. From an evolutionary perspective this makes total sense, but here we have to realize that this creature changed so greatly regarding its neck deviating from the creature it would have descended from to a point unlikely to occur. Here again the fossil remains need to be reevaluated without the guesswork by artists or those making assumptions when filling in the blanks from the found fossils that are often missing many parts.

Adaptive evolution obviously occurs. The question is what is the scope of change that can occur if any limits to the scope of change do actually exist?

Chapter 15

Like with Like

There are a few points about the Biblical account of creation when reading the older authoritative Bible versions that are undeniable. Too often people are inserting concepts or words or thoughts into the Genesis One text that simply are not there, which is mostly due to the more recent post-Reformation Bible versions mentioned in an earlier chapter.

These recent post-Reformation rewrites of the text are poor translation attempts at best, and outright lies at worst. These deceptive versions taint the text in a way that misleads or deceives the readers into seeing an entirely different and very inaccurate picture of the text's meaning. We can assume that there will be many people who when they learn that they have been deceived in this way by these very inaccurate Genesis One creation accounts, will be quite upset. This is likely to be especially true with those who, due to those very inaccurate translations, have ultimately rejected the idea of a discerning Creator God. Rational people, upon learning the truth about these lies, will likely reconsider the more accurate information found in the authoritative Bible versions, and thus rethink their position about the Bible's accuracy and maybe even about the existence of a discerning Creator.

So what kind of points does the Bible make that are, as I put it earlier, "undeniable"? What could possibly be undeniable in a book that is often condemned as "mere stories"?

What is a Kind?

What is a "kind"? As touched on a couple of times earlier from different angles, a "kind" is typically thought to be a specified group of creature that is particular in form and does not deviate from its fundamental form. From a creation perspective "kind" is a standard term. However, "kind" is typically rejected and is considered an unacceptable term in evolution circles. The term "species" is generally much more accepted by evolution supporters. But as mentioned earlier, from a translation standpoint, the two are equal since the Latin "species" used in the Vulgate has been translated to "kind" in both the Douay and King James versions of Genesis, which is very important to understand and remember.

For many years the term "species" was used in attempts to group the various theoretic creature groups on the evolution charts. However, as research progressed it became evermore difficult to categorize "species" in an acceptable manner due to the continuing variations found. Thus, the phylogenic model was created to account for the many deviations that did not nicely fit into the specific species classifications that existed at any point in time.

The modern phylogenic model does not really have species type classification because the belief is that it is an ever-changing ever-evolving process that has no end, and thus the changes are incremental and generally cannot be seen from one generation to the next. This is a fair enough approach, because if we cannot make specific distinctions via species definition then we best not make them at all.

Now regarding the term "kind" or its Latin equivalent "species", from a Biblical standpoint, we must examine specifically what the text actually says and what it does *not* say. There is this sort of unspoken thought that the Genesis creation account has a large listing of creature types or kinds, but if you were to ask a creation supporter, they would generally tell you there are no dogs or cats listed in the creation account. However they would likely go on to say that those are automatically included in the various kinds mentioned in Genesis.

This is where the real blind spot with creation supporters occurs. While they might say in words that that cats and dogs are not listed in the creation text, in many of their minds there is somewhat of a hard-set picture that subconsciously exists which includes cats and dogs etc. It is possible that this subconscious picture of cats and dogs and other such creature types is correct, but the truth is that such creature types, while certainly able to be included in the broad descriptions given, are simply *not* mentioned in the creation account. Creation supporters are not alone here. Evolution has similar subconscious issues with all of the artists' renderings and artificially completed fossils.

What is a "kind"? The Bible does not directly specifically define the word "kind" or "species" but it does make the undeniable claim that there are grouped "kinds". This neither confirms nor denies the possibility of evolution, adaptation, or static kinds that do not change much.

Douay English Genesis One:

"[20] God also said: Let the waters bring forth the creeping creature having life, and the fowl that may fly over the earth under the firmament of heaven. [21] And God created the great whales, and every living and moving creature, which the waters brought forth, according to their kinds, and every winged fowl according to its kind. And God saw that it was good. [22] And he blessed them, saying: Increase and multiply, and fill the waters of the sea: and let the birds be multiplied upon the earth. [23] And the evening and morning were the fifth day. [24] And God said: Let the earth bring forth the living creature in its kind, cattle and creeping things, and beasts of the earth, according to their kinds. And it was so done. [25] And God made the beasts of the earth according to their kinds, and cattle, and every thing that creepeth on the earth after its kind. And God saw that it was good."

During these two creation events listed as day five and day six, the "kinds" are very broadly stated, and judging from the phrasing used on the day three events regarding plants, the things that reproduce do so "after its kind". Based on the language used on days five and six specifically implying that creatures have offspring after their kind in the various Bible versions, we can safely infer that since several plant kinds are established on day three and that those kinds of plants do produce offspring "after its kind", we can safely extend this to any living thing.

Now, again it must be pointed out that this does not preclude or support evolution or adaption or static kinds, but rather it indicates as stated in an earlier part of this book, that when creatures or plant life have offspring that the offspring will be of the same configuration as the parent kind. This means that a dog will never give birth to a cat when two dogs breed, and the offspring will have the general characteristics or attributes that the parent dogs have. This is undeniably true, and on this particular subject the Bible is one hundred percent accurate and cannot be denied. "After its kind" says nothing about long-term changes through many subsequent generations, but rather that the immediate generation level offspring will match the parents' "kind".

Why Are Zebras Stripped?

When trying to understand the Biblical "kinds", people often take liberties with the textual interpretation and insist that it is all inclusive of all "species" that *ever* existed. Then when being challenged by evolutionists, creation supporters are typically taken to task regarding the various breeds of dogs, of which many can be considered unique species, but have been produced via human influence, thus they would be unlikely to have been created as they look today. This particular argument destroys the creation argument point made that "**all**" species were included at the time of creation. This situation regarding our ability to crossbreed animals to develop what can be categorized as different species is a very important point to consider regarding the evolution-versus-creation debate.

The fact that we can alter creatures through cross- and selective-breeding has profound implications on both sides of the debate. On the creation side it does confirm that the more articulate supporters of creation are at least correct about the ability for a creature to adapt or change to a limited extent. And this does seem to be a ***limited*** capability, in that no one has yet been able to breed them clearly out of the dog category as far as is known today. However, on the evolution side of the debate it is vague evidence that substantial changes can occur and that, in theory given enough time, the animal could morph outside of what we might consider a dog.

This brings us to other creatures that exist today that are very similar to another creature also still in existence. Let's take a look at *zebras* versus *horses*. These two creature types that are typically considered to be distinctly different species are extremely similar in structure and differ mostly in their outside appearance. We assume that they have a common ancestral lineage, but cannot prove this for certain. From an evolution perspective it is generally believed that horses were the likely the earlier progenitor. It is said that through many generations of offspring, horses that blended with trees and tall grasses survived better, thus those that blended better with the surrounding environment would be more prolific and therefore more readily survive, ultimately ending in the black and white striped creatures we today call "zebras". But this is not a logical explanation because it does not explain why they are specifically black and white. Since zebras are usually in a softer sandy or straw-colored environment, then based upon the environment adaption model of evolution, we would then expect that they would have evolved or adapted to have colors that more closely match the tall grass or trees of the environment. This would more than likely have caused them to be closer to the color of a lion or gazelle but with stripes, which they are not.

So regarding zebras, we then come to this point of asking if they could have been included in the initial creation account. Based upon the Genesis text it is a distinct possibility, however the text gives no indication whatsoever that these very specifically striped creatures are for sure included in the creation events on days five and six. We simply do not have enough information.

Beauty In Nature

There is great beauty in "nature" as most of us realize on a regular basis. Even Darwin pointed this out. But what is beauty? Who decides what beautiful is? Why do we humans look at something and have a general consensus that a particular creature is beautiful? There is definitely varying opinion in this regard, but the general consensus is fairly strong regarding what creatures are "beautiful" or what in nature is "beautiful".

The common expression, of "beauty is in the eye of the beholder" has a great amount of truth to it, but in the human realm it is more of a specific characteristic preference point, rather than humans over all. When we look at certain animals we often look at the species and will consider this or that species beautiful as a group. This consensus or stereotype is typically very strong. We have to wonder where this beauty preference comes from. Is it just that we see this on a regular basis and therefore are accustomed to it and thus that is what we like? That is the sort of opinion that typically accompanies the evolutionary perspective, and there is undeniable proof that it is at least partly true that we like what we are familiar with. However, when very young children are exposed to colorful birds that are not a part of their life or standard environment, even the very first time they see pictures of these unique birds, usually tropical in breed, they will be smitten with them and seem to find them attractive in that they will dwell on them.

We could say that the young children are attracted to these tropical birds because they are uniquely different than any other bird they have ever seen; however, the same is not true of less colorful birds that they may also never before have seen. This would disprove any argument implying that the young children are attracted to the colorful bird only because it is *different* than what they see in their normal environment.

There are clearly common points of attraction for us and in what we see as beauty, while it follows a general rule of thumb rather than being absolute, it is very consistent in our eyes. Beauty in nature is all around us and we have an affinity for things that are commonly considered beautiful, which seems to be natural in us humans.

Creature Groups

Touching once again on the creation's "kinds" issue, we will further examine the "kinds" listed in the creation accounts of day five and day six. As mentioned early on in this book, days five and six have a general distinction of egg laying creatures versus mammals. Now recall that the terminology of "whale" was done with creative license when they translated it to English.

While the text does not specifically indicate egg-laying creatures, it does seem to be dominated as such on the day five events. And the creatures on day six tend to be a better fit for the mammal classifications. Let's take a deeper look at the text and see what these groups more specifically are and if we can justifiably make the distinction of egg-laying creatures versus mammals. Here again is the text:

Douay English Genesis One:

"[20] God also said: Let the waters bring forth the creeping creature having life, and the fowl that may fly over the earth under the firmament of heaven.

[21] And God created the great whales, and every living and moving creature, which the waters brought forth, according to their kinds, and every winged fowl according to its kind. And God saw that it was good. [22] And he blessed them, saying: Increase and multiply, and fill the waters of the sea: and let the birds be multiplied upon the earth. [23] And the evening and morning were the fifth day. [24] And God said: Let the earth bring forth the living creature in its kind, cattle and creeping things, and beasts of the earth, according to their kinds. And it was so done. [25] And God made the beasts of the earth according to their kinds, and cattle, and every thing that creepeth on the earth after its kind. And God saw that it was good."

The waters brought forth:

- creeping creatures
- winged fowl that may fly
- living and moving creatures
- great whale (creatures)

The very broad categories just listed allow for a tremendous amount of versatility that could include mammals, but do not necessarily do so. Since the waters brought forth, we can make a reasonably safe assumption that many were water creatures along with other "great" or large creatures. The deeper you look into the translations of these various creation terms, the more consistent they are with science in each previous language that they were translated from. Our western culture, which dominates these topics, tends to use the English Bible translations, and thus we have typically lost sight of Hebrew, Greek, and Latin versions of the text. Because this book is not about translations, we are limiting the details in that regard. If you want more information about translations read the book *Understanding The Bible – The Bible How-To Manual* AND *The Things We Don't See*. Day five could easily include what can all be described as egg-laying creatures.

On day six the Earth brought forth:

- living creature
- cattle (Greek – tetrapous(4), Latin - jumenta)
- creeping things
- beasts of the earth

This is where the egg-laying creature versus mammal division struggles a bit. This is because it is a somewhat more difficult to fit in mammal whales into "let the earth bring forth" category. However, the "living creature" designation is broad enough for that to be the case, but the mammal fish would have to have been brought forth at the edge of the waters likely in a mucky earth area. I don't like to press this issue too much, but rather simply offer it up as a possibility in effort to get people thinking about that possibility so as to analyze its efficacy. It is important to note here that the etymology of "whale" is different than what we picture as a whale, so even that term is in question regarding our perception of it.

Together the full day 5 and day 6 list is as follows:

- creeping creatures
- winged fowl that may fly
- living and moving creatures
- great whale (creatures)
- living creatures
- cattle (Greek – tetrapous(4), Latin - jumenta)
- creeping things
- beasts of the earth

Make note that the Greek "tetrapous" implies four legged creature in general, and Latin "jumenta" implies beasts of burden which are typically four legged creatures. Both are connected to the English term "cattle", which is more of a statement of property than anything else, but the cattle that are property in a Biblical sense are typically four legged beasts of burden.

In the end, due to the fact that these Biblical terms are so broadly grouped, they cannot be denied being accurate. If we adhere strictly to the creation text as it is presented, it allows for tremendous versatility that is undeniably accurate to the generalization of creatures alive today and any fossils ever found.

As you can see in the creation text, some of the debates between creation supporters and evolution supporters are illegitimate due to the broad nature of the statements in the

Bible's creation text. Trying to nail someone's definition down beyond what the text says is not very fair, scientific, or logical. Biblical creation debates insisting that the Bible is making distinctions that include man's more recent *kinds*, *species*, or *phylogenic* classifications is a waste of time because the Bible says nothing of our manmade barriers of such designations. Depending upon how you break them up, the "kinds" designations in Genesis are limited to about eight in quantity if we keep days five and six separate with regard to "kinds" divisions. We tend to apply days five and six to larger type creatures like fish and birds and land animals, but they likely also include tiny sea and land creatures that we often refer to as bugs, worms, and amoeba, etc.

Chapter 16

The Elements of Life

What are the elements of life? How many elements are there? And how did they get there? These questions get very messy very quickly because we can keep drilling down into, and even beyond, the quantum level. So here we'll keep the questions a bit more basic and assume the quantum levels are well established and are ultimately what makes up the organic materials that are needed for biological life to occur and exist today.

What are the elements of life that are needed for the day five creatures to exist? At this point of analysis we sometimes encounter debate insisting that there was no death or decay prior to the fall of man in the Garden of Eden where Eve and then Adam ate the forbidden fruit. While the possibility does exist that there was no decay or death prior to the fall of man, it is unlikely that this was the case. One could argue that God instantly made all of the bacteria that is needed in the ground for plants to thrive, but the more likely path is that it took some time and that those plants and the gathered "waters" had to have first come about as a critical element of required life as witnessed today.

In order for the creatures listed on days five and six to be able to live they would most certainly have needed vegetation. Now while all of those creatures could originally have been herbivores, thus not eating other creatures, the possibility that they did immediately eat other creatures does exist.

Goldilocks Planet.

The position of Earth is without question in a "Goldilocks" area, but this is not the distance from the Sun alone, it also includes the rotation and orbit speed of Earth and the size of our Sun. It's not so much a "Goldilocks zone" as it is group of Goldilocks conditions overall that are "just right". It is these conditions that allow for the elements of creature life to thrive.

Air with its all-important oxygen, and bio-matter in the soil, heat from the Sun and Earth, and most importantly, the shelter and food supplied by the plants introduced on day three are all scientifically consistent with the needs of the days five and six creatures. If you have read *The Science Of God Volumes 1 and 2* you will have a good grasp on the order of events and how they worked together in preparation for the introduction of the days' five and six creatures. The so-called "Goldilocks" planet situation allowed for all of the required elements for creature life to thrive. Even many of the day five creatures are subject to the predatory nature of many day six creatures, such as bears catching fish in rivers, etc. Also, the predatory creatures created on day five could have thrived on other non-predatory creatures introduced on day five.

The Bible's creation account offers only very broad creation statements and it is for us to figure it all out. A great deal of debate about creation is due to people on both sides making incorrect assumptions about what the Bible *actually* says.

Relativism Abounds

So much of our perspective is relative to our understanding which is shaped by our previous knowledge that is derived from our upbringing. Our view of the creation and evolution models will be hindered to a relative level based upon the way we understand the text, which is relative to our education thereof. This is a somewhat inescapable problem we each have, and until we realize that we do this-or-that wrong thinking, it is a problem that will continue within any one of us. Once we personally come to terms with our limited amount of knowledge, it can set us on a path to truth, but our stiff-necked nature and our relative understanding of the Genesis One creation text causes us to stop searching and learning. We insist things have to be *our* way and regardless of which side we are standing on, we often get quite nasty about defending our position.

The level of belligerence is considerably higher on the evolutionist's side, mostly due to the fact the Christians are supposed to promote peace and love, or else they too would likely be equally belligerent about defending their positions. If you watch the level of resistance to either side of the theories, you will notice that, as a rule of thumb, when people have anomalies in their theories they tend to get relatively angrier or frustrated as they struggle to offer legitimate answers. And the more they have to struggle to offer answers, typically the further from truth they are moving, bringing us to the question of what truth is.

What is Truth?

It is striking how few people understand what "truth" is. When trying to describe "truth", people will often begin to explain by saying things that are true things versus things that are not true. But that is not "truth" in itself. Interestingly we have a natural inclination as to what truth is, yet we have a difficult time explaining or describing it to someone. Truth is what *is* and it does not change. Circumstances may change thus causing something to be true now or no longer true now, but the truth *function* has not changed and it never will.

Truth is one of those things that if someone doesn't know it then it is hard to explain it to them, yet everyone intuitively knows truth but often we do not realize that fact. If a young child lies to you and you ask them if they are lying and they sheepishly say "no", but you are certain that they are lying because you saw them do the act about which you are inquiring, it proves that they do understand truth. They very fact that they do not want to admit to their indiscretion clearly illustrates that they do not want to get caught because *they know that they are guilty* and do not want to face the humiliation that accompanies their ill deed. If they are capable of hiding the truth in that way, it proves that they recognize truth. This is a partial explanation as to why we might get angry when we are unable to answer the questions when our erred theories are taken to task. Those errors are then being exposed as inaccurate, which makes us guilty of perpetrating error. In a sense, we are lying to others as we promote our erred perspectives.

But, there is no need for us to act in this way if we simply offer our thoughts as possibilities for others to consider, rather than insisting that our way is the only way. Evolution and creation have, to a great extent, an ability to coexist without violating the fundamental premise of either theory set. But due to the stiff-necked rigid stances we take regarding our theories, we tend to hold to the inaccurate parts of those theories even tighter than we hold to the parts that actually make sense and are logically true–the chasm between the building blocks of both theologies not only continues, it thrives.

Basic Building Blocks

In the evolution models the primary building blocks are single-source origin, progressive change over long-periods of time, and no static divisions at all. The basic building blocks of the creation models commonly heard are multiple initial kinds, an ability to adapt, and static divisions that are never breached.

1. The first of each group *single-source* versus *multiple initial kinds* are absolutely opposed to each other.

2. The second set of building blocks of *progressive change over long-periods of time* versus *an ability to adapt* are nearly identical outside of the potentially large time-scale for evolution.

3. And on the final set of theory building blocks we have *absolutely no static divisions* versus *absolute static divisions*.

Starting from the last set of blocks, these two can be in agreement if we combine the second set of blocks with the third set. In this way a broad "kind" can deviate through adaption and would be allowed a great deal of variance due to the broad nature of the "kinds" listed on Genesis' days five and six. But the time-scale can be problematic depending upon which creation theory we are referring to, namely whether or not it's a six-twenty-four-hour day theory; if it is not, then the two can fit comfortably together.

It is now, and quite honestly always has been, the first set of building blocks of the two general theories where the contention exists. Evolution says it started from a single-source, where creation on the other hand says it started from multiple distinctly different sources. Evolution claims that because nearly all attributes of fundamental life are shared that they share a common ancestor, but this is a nonsensical argument. If a Creator created, then it is logical to assume that *common systems* were developed and employed from creature to creature. So this part of the argument comes down to whether or not the creatures were created in the forms of the broad kinds listed in Genesis One.

The evolutionary charts do tend to, very early on, roughly align with the very broad groupings laid out on days five and six in Genesis. It needs to be noted at this point that both sides practice a substantial amount of blind faith in both sides' various theories. Creation supporters typically take the Biblical account at face value, but unfortunately often will inadvertently or deliberately add a great deal of information that is simply not there when discussing the subject. The blind faith practiced by the evolution supporters occurs at most levels of the phylogenic or other evolutionary charts, especially those at the beginning of the chart. Now, it is claimed by the evolution side that the deeper you go in the geological strata then the more basic or primitive life appears. I will mostly steer clear of this in this Volume 3 of *The Science Of God* because it is discussed in-depth in *The Science Of God Volume 5 – Boats, Floods, and Noah – The Deluge*.

The issue that many people have regarding the geological column is that when it is thoroughly scrutinized you will find that it is not really based on anything absolute. It is a compilation of estimates gathered by various geologists and archeologists who have found interesting specimens or aspects of geology over the course of many years. These findings have been arranged to place them in chronological order based upon the *estimated* ages attributed to the findings. Much of this was done many decades ago back in the 1800s, and most age estimates given in modern times use those geological column age indices to estimate the ages of modern findings. Problems with the geological column run much deeper than one would imagine. Science is not always what it is promoted as.

Chapter 17

Do the Fossils Tell the Whole Story?

In the last chapter the idea that life gets simpler the lower on the Geological column you go was discussed. According to the geological column this is certainly true. But what if there are anomalies in the geological column? Could it be possible that the Geological column is not perfect? The Geological column is fairly old, and is hard to attribute to a specific time because of updates to it that have occurred over the years, but it is thought to be initiated during the mid-eighteen-hundreds era. In our modern era, the Geological column is not as needed in order to estimate the age of fossils because modern dating of fossils is often done through one of the various methods of radiometric dating. The Geological column is not specifically needed, although it might in many cases be somewhat more accurate regarding the order of events than is the wildly ranging various methods of radiometric dating. The Geological column does see a great deal of use regarding the general order of layers from the basic geological periods that it reflects.

The idea of evolution is that things began in a very simple form and became more complex with each subsequent generation. Over long periods of time the different creatures would eventually die and be preserved in the dirt. Being the humans that we are, it is intuitive to us that the layers of dirt and rock that we see are deposited in the order that they occurred, meaning that the lowest layers are the first or oldest, and the upper layers are more recent. This implies that any creature fossils found in the lower layers would predate or have come before those in the upper layers.

Groups of these layers make up the Geological column and have been named and are referred to as "*Cambrian, Ordovician, Silurian, Devonian, Carboniferous, Permian, Triassic, Jurassic, Cretaceous,* and *Tertiary*" periods. When a fossil is found in one of these groups of layers, then that fossil logically was trapped in the layer during the time period that the specific group of layers was deposited.

The current understanding of the fossils that are found is that those found in the earlier layers are simpler organisms, and the higher up the column the layers go, it is believed that the fossils then are more complex and newer. This logic is difficult for creation theories to combat. This can get a bit tricky in debates and it comes down to the time-frames attributed to the various geological layers' periods. What we don't know is exactly how old each layer is or how much time passed cumulatively from the lower to the upper. This is an area of great disagreement between evolutionists and various creation theories.

The creatures appeared quite suddenly in a sort of creature explosion within a fairly limited portion of the entire column. However, since the stated timespans of the periods are quite long and the fossil finds tend to back up the idea of simple forms lower and more complex forms higher in the column, it appears at first glance to support the evolution model as it was designed to do. Yet, evolution still has one weakness in this column, and that, as mentioned before, is that if any far more complex creature is *ever* found below the simpler creatures in the current model, it completely invalidates the idea of simple-to-more-complex creature that is based on those layers.

If creation supporters are ever able to defeat the idea of single-source long-age evolution, it will likely occur through a trove of fossils found that are undeniably equal to or predate the simpler creature forms in the current geological record model. At this point, the evolution side of the debate seems to be winning that part of the debate. But we have to question the motives of many researchers. There are many people that support evolution and are utterly and often angrily opposed to anything to do with a Creator or the idea of guided creation. Many of these people get paid to further their theories. In doing so they are searching for very specific fossils in very specific geological period layers.

We usually find what we look for, but we generally do not notice what we are *not* looking for. If someone found a trove of fossils that is not in line with their mental model, then they will typically disregard it as a more recent fossil find and thus ignore it. This could be done to protect their financial interests and livelihood, but more likely, it would simply be misunderstood as newer, and thus will be taken as somewhat meaningless to them even though it may have actually been older.

Unfortunately for the creation crowd they are not as proficient as extracting money from the government for archeological and geological work. Some of such research is done via government grants to colleges that perform these archeological digs. Science grants are not as readily available for anything that can be construed as "religious". This means that if the grant is intended in any way to look for anything that might support creation, and thus the Bible, it is far less likely to get any substantial funding. However, if creation researchers could frame their position in a more intelligible manner and were a bit more persistent and ambitious in this regard, they could likely procure far greater funding to look for what they need to find in order to help further their perspectives. Remember, the evolution model is built heavily upon the idea that the lower in layers that life is found, then the older it is and it will be much more primitive or simple in nature. So a single complex creature found in a layer that is verifiably equal to or older/lower than the simple fossils found that support the evolution theories would utterly destroy the current evolution model.

Tree of Life

"The tree of life" non-Biblically speaking is the assembly of creatures beginning from the simplest fossils of the past to the most complex life forms that exist today.

Some of the simpler creatures that exist today can be inserted into some of the earlier branches on the tree of life if they have retained the same form through the ages. The evolution chart or tree is basically a tree of life. This tree is a valuable research tool, but a great deal of it is based on a single person's assessment of a connection, meaning that in most cases a single person made a particular decision to make a particular connection. However, these connections are made by many different people along the way. It's just that there is typically not a central scientific board that reviews every connection and then deems it legitimate. The connections are subsequently viewed by other people, and if someone should find an anomaly, they then might attempt to bring it to the attention of the person who made that connection if that person can be found and is still alive.

The problem with this tree is that it is not a centralized work and has somewhat arbitrary origins. You will see many renditions of the evolution charts created by ambitious researchers, students, and science departments in schools. They do agree on many connections, but there are differences in them. I only mention this to make people aware that the evolution flow of the "tree of life" is not an authoritative instrument; it is a conglomeration of work from miscellaneous people transferred from one person or group to the next, and it itself has a sort of tree path to follow if you want to get down to its own creation origins. While it might be a pretty close representation of the actual trail of life, it is largely speculative. We should not utterly discount it, but neither should we blindly accept it as "absolute fact".

Mono Phylogeny

Mentioned earlier in the brief list of terms commonly associated with evolution study is "taxon". Taxon is a group of creatures that has a specific taxonomic name. Those groups can be as finely defined as those naming them choose to define them. With the phylogenic classifications, there is ultimately no end because one could include every living thing, if that were possible. The resolution granularity of changes can be quite small if so desired. However, usually specific names are not assigned to a taxon unless it is warranted, but defining and naming is a judgement call that must be made by someone at some point, so again the naming and classification is somewhat arbitrary.

When a taxon is considered monophyletic it means that it descends from a single ancestor. Now this gets complicated very quickly and in order for it to continue very far along, creatures that reproduce with male/female sets must breed only with their siblings to remain "monophyletic". If it is *not* monophyletic, then at the most basic level it was an instance of crossbreeding or breeding with a first cousin or a creature more distant in relationship.

The classifications of the creatures has become a bit complicated to understand due to all of the various terms introduced in effort to explain the various attributes, relationships, and forms. When considering this along with the Greek- and Latin-ized naming of any previously- or newly-found fossils, it makes the whole business a bit daunting to dive into.

Don't Be Fooled

Between the details of all of the names and the jargon of the evolution industry, we can add to that the forensic artists' rendering of missing parts *not* actually found in the fossils, and then again add to that the pictorial artists' renderings of the forensics artists' work that is based upon a fossil that is often considerably distorted in form due to having been pressed between layers of heavy sediment for many years. This can produce a potentially very different picture than what the creature actually may have looked like. This is immediately obvious when you see the renderings of the same fossils from different artists. Some of these renderings can vary quite a bit.

Don't be fooled by any of this, take it for what it is. These are all speculation and best guesses. Do not discount them because there is nothing wrong with such speculation—until they demand that this is what is, or that it is "factual". As long as it is known that these are all *speculative* presentations there is no problem and thus people can then realize that they themselves could also speculate on these creatures and create their own vision of any creature in question.

It is all speculation and it is subject to change, but those changes when needed are slow and difficult in coming because much of this work is presented in school textbooks for students of all ages. Those who create the textbooks are more concerned about profit than accuracy, and thus such known errors often continue in revision after revision of these textbooks that are seldom corrected on such trivial matters. And in many cases the textbook companies' staff might not be privy to the deep details of evolution and might rely upon the schools to notify them of needed corrections, thus if no one informs them, they won't know that they need to make the corrections.

Reclassifying Skeletal Finds

As noted in previous sections, species or phylogenic classification is somewhat of an arbitrary task. At some point some person must decide where something belongs in the tree of life. The problem with this is that often the fossils lack a great deal of information, meaning that the greater part of them is not there—it is simply missing. In this very common situation, these creatures' fossils are compared to known full fossil or skeletal remains that best match the portion that is actually found of the fossil. From that they will make an assessment, right or wrong, about the fossil and proceed to place it in the tree of life, often to fill in a gap or area that is lacking in definition. Most people would be surprised at the limited and fragment remains collected in fossil finds that are often allowed a classification or place in the tree where the obvious gaps are found.

When a fossil or skeletal find is found to be in an incorrect position, it is difficult to remove that error from the charts and thus from culture due to the prolific nature of modern printing of textbooks and the passing along of information electronically. This problem was especially prevalent starting in the early part of the twenty-first century and became much worse as time progressed. At that time many people would create personal electronic or internet pages or internet sites and would then parrot these inaccurate charts, with no specific intention of updating for future corrections. In fact, the evolution indoctrination is so heavy that most people do not realize that these charts may change somewhat as research continues to correct the many anomalies in them. Which is the cause for people repeating and spreading errors that are built into the charts and theories that those charts are partially derived from. This is not true of the entire evolution model, but it is important for people to understand that this problem exists and persists.

The creation side of the topic does face a similar, but far less pronounced, version of this problem. Creation theories generally remain fairly constant in this regard because they are based upon an unchanging book that we call "The Bible". So the theories don't really change a great deal in this case, but there are errors that won't depart and keep being reiterated and passed along from person to person, namely the far reaching six-twenty-four-hour day creation model derived from the late post-Reformation Bible versions. It keeps getting passed along due to children being brought up on and taught from the aforementioned more recent post-Reformation Bible versions beginning in the seventeen- and eighteen-hundreds from which many Children's Bibles are derived.

We all need to understand that the so-called "facts" change. While we must use caution adopting these changes, they must be fully considered and taken seriously and introduced into public view, especially when they are correcting outright fraud or obvious errors.

Chapter 18

What Do You Really Want to Know

When people begin to research the creation-versus-evolution topic it is often initiated by a desire to know whether or not a God or Creator actually exists. That is a fair motive, but it's really important for *you* yourself to understand exactly *what* it is that you are trying to understand. Do you want to know if God exists, or are you trying to figure out how everything came to be regardless of the existence of a Creator?

A great deal of the Genesis creation account can be considered to be accurate and completely in line with many aspects of modern science. As you have likely noticed so far in this volume or if you read volumes 1 and 2 of *The Science Of God*, the Genesis order-of-events do agree with science on a scientific level. But eventually you will always come to a point where you have to make a decision of how *you* believe certain things were initiated.

You must understand that the initial points of origin in both big bang and evolution theologies are *speculative* no matter how viciously people insist that you agree with them and their "facts".

I am not saying here that those theories are incorrect, but rather that they simply are not "fact". Both the initiation of big bang and the initiation of evolution are <u>speculative</u> and anyone who demands otherwise is a liar or is sorely misguided. Speculation is fine, but unless we or anyone we know has personally witnessed it, we can only <u>speculate</u> as to how life got here and how life began.

Did Aliens Bring Life Fully Formed to Earth?

When discussing creature origin topics with some people, you will come across a grasping theory that aliens brought life to Earth. Even some of the most prominent pop-science talking heads will promote this ridiculous theory. We can consider this in regard to creature origin, and if intelligent life in other solar systems does exist, the possibility of aliens bringing life to Earth actually does exist however unlikely it may be. But from a scientific standpoint, the life does not need to be intelligent. Imagine a planet teaming with life and eventually a massive meteor crashes into that planet and breaks it into nothing but space debris. To continue the theory, some of that space debris possibly still contains viable life in the form of microorganisms. This somewhat unlikely scenario is nonetheless theoretically possible.

Let us imagine that single-source long-age evolution is an accurate understanding of creature origins. Then the "panspermia" theory, as it is called, could work where microbial life came from another planet. A pre-established set of instructions is carried from this destroyed planet through space and eventually encounters Earth. As it enters the atmosphere it breaks up and the microorganisms are spread abroad and evolution of those microorganisms begins, and then hundreds of millions of years later, here we are! Problem solved!.. Except for the fact that we are trying to specifically *understand how the first life began*, and that theory does not tell us or explain that in any way. In this also very unlikely theory, life from the other planet started life on Earth, but the theory fails to answer how life began on the starting planet in that theory. Kicking the can down the street in this way does nothing to answer the ultimate question of, how did life begin regardless of where it began?

These nonsensical theories, while possible, are highly unlikely. Even the intelligent aliens bringing life to Earth in a spaceship still does not answer how those aliens would have come to be. These theories are somewhat childish in the respect that they simply do not answer the question of how life began regardless of where it began.

To add a few thoughts to the "panspermia" theory; the likelihood that life is going to survive the hostile radiation and severe cold of space, not to mention the devastating heat and radiation from the initial destructive impact that would have sent it on its way to Earth, and the atmospheric entry-heat and finally the impact and heat created as it impacts Earth, life is unlikely to have begun that way on Earth. Similarly, if we are anywhere near correct in our time and distance estimates of even the nearest star to our solar system, an alien would have to travel for several years at the speed of light to get from that solar system to ours. Imagine yourself on a spaceship for roughly four years. Now if you are traveling at the speed of light and a tiny piece of meteor is floating in space, how will you see it soon enough to avoid it at the speed of light? The impact to your spaceship could potentially be devastating! You will have to deal with this issue for several years as you travel to even the *nearest* star.

Let's not waste our time kicking the initial point of origin down the street in this way, let us work to understand how life likely began *regardless* of where it began. In the end, we may never know for sure exactly how life began, but we can logic our way through a great deal of the information and define the points of consensus and the points of mismatch between creation and actual science. Once those are established they can be further studied to see if they are logical and reasonable. An additional interesting note here is that the Genesis creation account does not limit life to Earth alone. In fact, if the text is read and taken in the order it has been offered to us, then the creation account is all inclusive of the entirety of the Universe and the entire Universe likely includes every aspect of at minimum days 1 through 5.

Murchison Meteorite

In the year 1969 a meteor was witnessed as it entered Earth's atmosphere. The meteor was large enough that some of it remained after it impacted the Earth. This Meteor landed near Murchison, Victoria, Australia and thus is referred to as the "Murchison Meteorite". Analysis of this meteor indicates that it contains amino acids, which, as previously mentioned, are believed to be fundamental building blocks of life.

There are a few possibilities regarding this meteorite. The first is that it is from another celestial body that once likely had water, which when impacted had a fragment be projected out and the wandering fragment eventually made it to Earth in the form of the Murchison Meteorite. Another and potentially more likely theory is that the meteorite actually came from Earth itself as a result of a different meteor impact and orbited the Sun along with the Earth for a long time, then eventually it was reunited with Earth in the form of the Murchison Meteorite. The theory of Earth origin is disregarded by science because the meteor is deemed to have an age of about seven billion years, where the Earth is estimated to only be about half that age, thus excluding Earth-origin as a possibility in that perspective. The estimated age of the Murchison Meteorite may suffer from selective believing and selective listening.

Another origin possibility is that is that the Murchison Meteorite is indicating to us that the Universe is teaming with life, so finding amino acids that we call "the building blocks of life" would be extremely common as alluded to in *The Science Of God Volume 2 – Day Three – Gravity, Land, Seas, and Evolution* of Plants. If a Creator does exist, then it is highly likely that life similar to that of Earth abounds throughout the heavens.

When studying specimens such as the Murchison Meteorite, we must protect ourselves from our previous indoctrination and clear our heads. We must evaluate the specimen and the data derived from that evaluation from a clear-minded and unbiased position. Selective analysis is the single most influential flaw of long-age evolution study.

Is the Murchison Meteorite the type of vehicle that brought life to Earth? Possibly, but as mentioned before that still does not answer how the building blocks of life are contained in the meteorite. We can assume that it came from another far away planet, but as mentioned earlier, that does not explain how life came to be on that theoretical origin planet.

No One Doubts

Earlier, the idea of adaption was discussed along with survival of the fittest theory. The theory is that creature patterns and colors were established through adaption and/or their survival ability due to their close match to the coloring of the surrounding environment in which they lived. Over generations those that matched more closely would be able to reproduce due to the fact that they survived and were not killed by predators. This would eventually cause them to more and more match or blend in with their environment. There are few people who would deny that this potentially could occur over time, but we are still faced with those creatures that don't fit that evolution model. Adaption or survival of the fittest in this way is a big part of Darwin's key theory about evolution.

There are other ways to interpret the patterns and color of creatures, which is most apparent when a creature does **not** fit with the general idea of color and pattern changes due to the survival of the fittest. If godless evolution is truly how it occurred, then not everything needs to have a cause. Some patterns could spontaneously or randomly occur. If we can imagine that random lightning stuck some gases or chemicals and created amino acids and then from that all life arose, we can certainly imagine that random colors and patterns could occur. The evolutionary laws are *speculation*, **not** fact.

The Laws of Science

Since "science" is a quest to "know" it might seem logical that there would be scientific laws, but it's not that simple. "Laws" are man-made rules that we apply to things to test them. In saying man-made I mean that the definitions are man-made.

Science should never really proclaim "laws" because the idea of unchangeable laws tends to stunt the growth of the minds of upcoming scientists. We see this with the evolution theory that was forced onto society as if it was "undeniable immutable fact", which is simply *not* true. When children grow up with the idea that evolution, as presented, is true and it is firmly cemented in their minds, rather than them understanding it as the speculation that it actually is, it then causes them to not understand that there are other possibilities to look for and research. This in turn cheats them out of researching for themselves and cheats them out of a more robust learning experience because of the propagation of the myth of "laws".

When it comes to the "laws" of evolution, or the "laws of physics" for that matter, you will often see or hear people explaining how these laws dictate the outcomes. But since there is a great deal that we simply do not know about what we study, we must realize that the laws do not cause what occurs in nature, but rather that our man-made laws are dictated by what occurs in nature. "Laws" are the result of nature, not the cause of it. Without particles, the laws of physics would not occur. And without the functions of biology, creatures would not exist. The functions dictate the laws, and to the best of our ability we examine those functions and try to record them in the form of "laws" to help us better understand consistencies that we observe. Nature works, and the functions we see are what we derive our scientific "laws" from.

Chapter 19

Everything Has Been Saved for You

The geological record is a true treasure trove of data for everyone to study and enjoy. With this information we can make great assessments about the geological history of the planet on which we all live. There are great tales of ages gone by to be told by the sediment, but how we interpret that information is what we all must work to do properly. There are many debates and questions regarding the ages of the layers of sediment, and those debates and questions are in addition to of all of the creation specific questions those layers invoke.

Many of the tales told by the soil of this Earth are physically written in stone and are unchangeable. What we *think* the Earth is telling us is absolutely irrelevant to the truth of what the Earth *actually is* telling us. But are we willing to listen to the Earth's geological tale? Some people think the layers have been deposited over millions and even billions of years as the result of natural geologic movement and weather etc. Some people think the layers reflect a six-twenty-four-hour creation period. Others believe the layers were deposited as a result of creation over millions or billions of years. And some people believe that sediment layers are from a worldwide flood that occurred during recorded human history, the inaccuracies and potential realities of this are discussed in *The Science Of God Volume 5 – Boats, Floods, and Noah – The Deluge.* Are any of these a possibility? Are any realistic?

Selective Listening versus Selective Believing

As we venture into research and study into the creation-versus-evolution logic-struggle, we are faced with a fairly wide selection of choices as to what to believe. This in itself is a problem in that we should not be choosing what we believe; rather, we should be examining the information and then analyzing it, and from that analysis come to *fundamental* conclusions, rather than *final* conclusions. We can get philosophical and question things on a more obscure level, but in this book we are trying to keep it tangible and rational. Our selective *believing* is almost always predicated on our upbringing and on our education. Both our upbringing and our education are not rational thinking sources, but are more in line with *indoctrination* rather than with *reason*. There is nothing wrong with the indoctrination side of receiving information until we blindly repeat that information in effort to "teach" others. If we have chosen to selectively believe in that which we were indoctrinated about and then go on to share that information with others, if the indoctrination information we chose to believe happens to be wrong or incorrect, then when we share that information, are we actually teaching someone? Or are we lying to or inadvertently deceiving them?

Selective believing is an age-old problem and is typically thought to be a religious/creation side problem, but this is not true. Selective belief issues from indoctrination are far more prevalent in the evolutionary sciences than in Biblical circles. In most colleges, even some Bible colleges, Darwinian evolution is taught as "fact" and it is not discussed in a rational manner where the student is allowed to reason through all of the information. Instead they are indoctrinated with a pre-selected set of information and decades and even centuries old theories, and they are generally not allowed to openly hear and fully analyze any other information that might fall outside of the evolution tales told by their indoctrinating "educators".

When the indoctrination of selective believing has come to maturity in the mind of the receiver, then selective listening begins to do its ugly work to stop otherwise open minds, causing those minds to slam tightly shut and ignore any facts that are not in alignment with their aforesaid indoctrination. This problem seems to have increased and decreased over the centuries, but hit epidemic proportions at the end of the twentieth century and was multiplied many times over at the beginning of the twenty-first century to a point of suppression of ideas and utter censorship of any opposing information. The take-away of this is that we all need to open our minds and hear and consider other perspectives and then do proper analysis of that information. For most people the resulting understanding will be far different than what they currently have selected to believe. If only they could make judgements on actual fact alone.

Believing Wrongly

When we cling too tightly to our selective listening and/or to our selective believing, we make ourselves susceptible to believing wrongly. This blind faith that we practice is perhaps the most prominent obstacle we humans face in trying to understand our origins. When we adhere to the predetermined interpretations of either evolution or of Biblical creation without us unbiasedly evaluating the information for ourselves, we will inevitably eventually believe what is either inaccurate interpretation or outright lies regarding this subject.

When it comes to the Biblical creation account, it is our lack of knowledge about the various Bible translations and incorrect beliefs about what the Bible is and what Genesis is communicating to us that causes our understanding problems. On the evolution end of the origins subject we have to deal with the pre-suppositions established decades and even hundreds of years ago regarding the order of events and the estimated ages of fossils found, along with the estimated age of any geological layer in which any one fossil was uncovered. In addition to that we have great potential error in the assumed correlation of the geological layer order where fossils were found at one location on Earth, relative to the layer order where similar fossils were found at another location on Earth.

When you study into, not evolution itself, but rather into everything that single-source long-age evolution is based upon, you will find that it is not exactly as solidified as is supposed and as it is presented to us. The mountain of evidence that we are presented with regarding evolution is thought to be far greater than the amount of evidence that we are presented with regarding creation.

A Mountain of Evidence

We are presented with mountains of evidence about life that occurred many years ago on Earth, but there is a battle regarding the meaning of those mountains of evidence. What is the land trying to tell us? And, are we open enough in our thinking and rationale to accurately interpret those mountains? There is a great deal of confusion that occurs in the minds of many creation supporters where they tend to accept certain scientific suppositions and then try to explain the creation model with those suppositions incorporated into their theories. Most of this has to do with the geological record and the "geological column" and the suggested ages of the layers or periods therein. When these creation theories are scrutinized they typically fail such scrutiny because they blindly accept certain points of "scientific" reference as if those points are absolute.

The most prominent trap creationists get themselves caught in is when they accept long-age creation and then must explain how the mountains could possibly get covered by water as the Genesis flood account indicates. In the Genesis Six and Seven global flood account of Noah's time, it says that the highest mountains where covered by fifteen cubits depth when the water finally stopped coming. Could this be true? There is little argument in their minds as they take this route, but they are frequently tripped up and defeated due to this particular rationale. If you want some additional insight on this particular issue read *The Science Of God Volume 5 – Boats, Floods, and Noah – The Deluge*. In it you will find some details that are typically unknown or unrealized by most people.

In addition to the issues surrounding the mountains being covered by water, there have been fish fossils and shark teeth fossils found at high altitude levels implying that the mountains were once under water. However, if you are a creationist don't get too excited because you will immediately be told that over hundreds of millions of years or some obscurely long period of time that the mountains had once been at the bottom of a lake or sea basin and that over a long period of time, the continental drift caused a buckling of the land and thus forced the land up to high altitudes bringing the shark's tooth along with it. This seems to be a somewhat logical explanation of how such fossils could get to that high of an altitude, but it is dependent on many presuppositions that again are discussed in depth in *The Science Of God Volume 5– Boats, Floods, and Noah – The Deluge*.

If a shark tooth fossil is found, then it is a shark's tooth fossil, and we can all accept that, but if that tooth is found and is not accompanied by the remainder of the shark, then we might have a problem of overreach. If a shark tooth fossil is accompanied by much of the rest of the giant shark then we have something to sink our shark's teeth into. If not, then well... then it's just a shark's tooth and so we will just leave that thought hanging.

There is a great deal of evidence in regard to the mountains and the formation thereof and the fossils that they contain. This evidence tells us all a great deal, but our selective listening inhibits us from understanding the truths that these mountains of evidence are telling us. While the mountains do hold a great deal of evidence, they are of little to help us in understanding the initial points of creature creation or their arrival. The creatures or creature parts that are found at such high altitudes are fully formed and are *not* "primitive" life forms, so we must completely discount them in the origins debate.

However, there is a great deal of information in these mountains if you apply true logic to it all. But watch out for the moving targets as you do so, because every couple of decades timelines get adjusted to compensate for logical inconsistencies found in the prominent geological theories.

Chapter 20

Is There Any Meaning to It All?

With all of the evidence found, and there is much, what does it all mean? Is there any meaning to any of it at all? When you review the fossil finds of the past couple of hundred years, you will be overwhelmed by the vast amounts of fossils. In fact, there are so many that it is unlikely the any one person could ever study them all. You would have to travel the world and dig through boxes and drawers of archival finds that are not on display anywhere. In addition to that you would have to visit every museum and college that has a fossil, of which there are many around the world.

Most of what is presented to us about evolution or creation is found in museums, the data from those finds is then incorporated into textbooks and documentary programs for our viewing consumption. With all of this evidence of evolution how could it possibly be wrong? So much proof must certainly prove beyond any doubt that single-source long-age evolution is certain and that Darwin was right... Right?

What is "proof"

If you discuss the origins topics with creation supporters you will usually end up with the Bible itself being the proof that the Creator did it all. In a religious sort of way that is a fair view, but it is scientifically lacking. After all, if there is a Creator that did in fact create, then we can assume that there would be some evidence of that creation process. The Bible is simply not proof of creation, but rather is an account of creation, or in this discussion, a theory of creation for which we must find proof. The problem with this is that the godless evolution perspectives have all hijacked the evidence and adjusted the conclusions to match evolution theories. Creationists are usually defeated in this area because they do a poor job presenting their case, usually due to some blatant and glaring logical inconsistencies and then when they get backed into a corner will typically revert to "The Bible says so, so God did it all." Not that they are wrong in that, but it is an insufficient argument to anyone who supports the godless-evolution perspectives. And in the case of creationists reverting to the "The Bible says so" argument, I have to agree with the evolutionists where they say that taking that position is a fool's argument, even if that belief is ultimately correct.

A real God or real Creator is not going to ask us to believe lies and will not do anything to deceive us. A "Loving God" would not lie to us. So with that, if you believe the Bible, then you should not need to use the Bible as your proof, but rather use the Bible as a springboard to find the evidence that will prove the Bible's creation account to be an accurate account of creation. In other words, if that is the way it occurred then there will be evidence to support that creation account.

But now we get to the evolutionist argument and the accuracy of *their* claims. I took a deep dive into trying to find a central authoritative point of reference regarding evolution and I found that evolution is both highly centralized and not centralized at all. You might wonder how this logical contradiction is at all possible. Well, that all depends upon who you are talking to.

There are many evolutionists who adhere to a centralized Darwinian perspective with a death-grip that cannot be broken. But some of these adherents will lose their grip when taken to task on some of the Darwinian conclusions, at which point evolution instantly becomes decentralized and it is suddenly said that "evidence can be found in any eighth-grade classroom science book". This makes it a theology of consensus. I have been given a list of great books that are about evolution, recommended by various people, and the books were suggested to be used as key sources for evolution information. But all of these books share the same inaccuracies and biases and offer no sound source references other than these and other similar books. The references that are offered are a combination of biological evolution and Darwinian evolution reference sources.

When observable-evolution in the biological sciences is cleverly or blindly blended with Darwinian evolution, it creates a very compelling presentation that lures many people into its snares. These books are not proof of much beyond simple observable biological evolution, which is little more than adaption or mutilation of micro-organisms that often are unstable and quickly die due to the violation of their general purpose and makeup when they "evolve" through intelligent laboratory manipulation.

It is far worse to imply in any way that the scientific proof for evolution is found in any book anywhere, than it is to claim the Bible as proof of creation. When referencing the Bible for the creation account, at least it is an undeniably old and consistent account of creation, and it is the only scientifically sensible account found in ancient writings about creation accounts worldwide. Most of the ancient written creation accounts loosely follow the Bible's creation account but are typically not very scientific. When taken at face value, the Bible's brief and broadly stated creation account does not violate science, whereas books on evolutions utilize circular referencing of each other as their verification.

Getting back to our tadpole/pollywog example from an earlier chapter, the tadpole idea is a more logical theory to explain fish fossils that have legs than is evolution. In other words, if a small creature can morph from a fishlike creature into of frog in its lifetime, then why not a larger similar creature since we find many extraordinarily large creature fossils similar to much smaller creatures that exist today?

There is a sequence that most of us fail to see in all of this, which is that many of the early theories about creation that have been either morphed into godless evolution or hijacked by "science" were initially offered by creationists of old that were attempting to scientifically explain creation and how God may have done it all. But these theories blew up in their faces because they stepped far outside of the bounds of the Genesis creation text and thus gave life to evolution in a way that they never imagined would occur. That is what happens when you promote inaccurate claims as if they are fact and truth. Most of these dedicated people recognized the order found in creation and were attempting to better explain it all. We can all confidently assume that their theories or conclusions would differ greatly if they had the data from our modern observations. What many of the creationists of old viewed as points of order, pop-evolutionists today see as somewhat random chance.

Order is Random

What is "order"? And, can "order" be random? In our logical human minds, order is a way of stating that things occur in a somewhat organized and therefore predictable manner. The order that was realized by past creation thinkers was clear evidence to them of a Creator who created as per the Genesis creation account in the Bible. But since some of these creation thinkers arose during the proliferation of the recent post-

Reformation Bibles spoken of in earlier chapters, they were influenced by some of those inaccuracies. This caused them to attempt to make clear distinctions of kinds that stepped far beyond what the Genesis creation account actually states.

Their error in trying to force definition into the interpretation of the creation text where no definition in the text was made greatly compromised their thinking and therefore their resulting theories. They wanted everything in absolutes because they saw God the Creator as an ordered entity that did everything to perfection.

What they failed to understand is that order can have many levels and because of the vast ability for the building blocks of creation to be ordered and rearranged, "random" creation can occur. Order can occur in a random manner and still be a design with order. Thus, we can have some level of evolution or change over time, but it will be limited and stay within the confines of the parameters of its point of order.

When discussing the idea of *random* and chance of evolution starting from a spontaneous event, the odds are thought to be extremely low, at which point the idea of infinity is often invoked with ridiculous examples. One such example for evolution to occur is that if you had monkeys clicking away on typewriter keyboards typing an infinite amount of letters, eventually one of those monkeys would, by pure chance, produce the full text of Shakespeare's *Hamlet* without errors. While this appears to work out mathematically speaking, the likelihood of it ever occurring in reality is zero. Infinity never ends, and thus the monkeys would be destined to type for eternity until one of them finally stumbled upon the perfect sequence. But since the mathematical combinations of letters on the keyboard multiplied by that number for each letter in *Hamlet* is so very large it is unlikely that it would ever occur. Now, since we are speaking in hypotheticals with this hypothetical example, we will venture a bit further down this nonsensical path to point out that one single duplication of one letter in one word that would have otherwise made the *Hamlet* text perfectly typed as inferred in the monkey-theory, would now force those poor monkeys to endure **infinity** all over again. This is clearly animal abuse. Poor monkeys!

So to the credit of those who support or buy into the monkey theory, they have given creation supporters a great advantage in logical reasoning. This is because random chance of evolution is equated with the unlikely chance that a monkey could ever randomly produce *Hamlet*, which is to say that evolution would take an infinite amount of time equal to the infinite amount of time for the monkey to randomly type *Hamlet*. Which in this case is basically illustrating that single-source spontaneous evolution did not occur. And since the age of the Earth is believed to be in the neighborhood of only four billion years in age, it is nowhere near infinity in age. So mathematically speaking, somewhere out there is an incredibly lucky monkey, to a point where it is simply illogical to even consider.

Superficial Hypothesis

Don't fall for such superficial arguments as monkeys typing an infinite string of characters hoping they will stumble upon having typed out, perfectly by chance, Shakespeare's entire *Hamlet* in an infinite amount of time. Accepting these superficial and nonsensical arguments are snares set by clever debaters to trip you up. If evolution did occur we should not need to make up fantastical imaginings about monkeys infinitely typing. And, let us not forget that monkeys **do** display some level of intelligence and intent, inanimate objects and chemicals do not. Another superficial thing you will see, mostly done by evolution supporters, is that they will come against a brilliant creation debater and when they run out of logical points to make, they will then use ad hominem attacks on their opponents. Creation supporters, if they happen to be practicing Christians, are not supposed to partake in unfounded personal attacks in that way, otherwise we can assume they too would practice this unsavory tactic.

Superficial arguments and superficial hypothesis are common in the evolution versus creation chasm. Once we dispose of the pettiness that often occurs in these debates, we must address the superficial hypotheses that are common on both sides of the debate. Even Darwin himself admitted that his hypothesis was superficial. While his view did evolve along with the creatures in his hypothesis, he at least was willing to admit that he was not sure about it all, though he did feel that his case was compelling.

Look beneath the surface on everything about creation and everything about evolution and you are bound to find some gems of information that most people don't see, or, more likely, refuse to see. There is a world of information yet undiscovered just waiting for **you** to find it, but you will not find it if you insist on proving one case or the other with flawed logic. Follow the actual *findings*, rather than people's *conclusions*.

Darwinism

Since the enlightenment era, a new religion of science has been born. Some science is good and true, while other science is built upon utter speculation and extrapolation of that speculation. When it comes to evolution, Darwin is the god of that speculation, and the religion is Darwinism or evolutionism. One could debate that Darwinism is not a religion, but then I would insist that they *publicly* define religion and explain how Darwinism and religion differ in front of friend and foe alike.

Chapter 21

The Darwin Delusion

The creation versus evolution debate should not be us versus them, but, sadly, it is. Instead of being able to calmly and rationally discuss this topic to drill down deep enough through the layers of pre-suppositions, we demand that our own position is the only way, and that our holy book is *the* holy book. In this case I am speaking of the holy book of Darwinism. This Darwin delusion has worked its way into the hearts and minds of far too many people to a point that they have become unreasonable and will refuse to consider any other points of view. You can try to explain something to them, but they will, without explanation, dismiss any opposing thought.

If you are going to dismiss someone's thought in a dialog about the evolution-versus-creation topic, you should offer a solid reason why you have done so. If you are unable to produce a coherent position against their opposing suggestion, then might it be that your theory is flawed? Indeed it is.

Animal Intellect

A big part of our human blunder and namely Darwin's blunder, is our general assumption that animals are stupid and without logical sense. At this someone might point out that Darwin is exempt from this foolish line of thinking because it is he who marveled at the fact that animals seem to *think* and make choices, as he expounded on in the relaying of his hunting dog experience. In his experience, a hunting dog went to retrieve two birds, the first was still alive and flapping around while in the grasp of the dog's mouth. As the dog came to the second bird to pick it up, it paused as if to ponder the situation, at which point the dog violently shook the bird so as to kill it and then proceeded to pick up the second bird together with the first in the dog's mouth. Darwin thought that the dog realized that if it had attempted to pick up the second bird then the first bird would have likely escaped the grasp of the dog. Thus, Darwin suspected that the dog *reasoned* through the situation and realized that it had to kill the first bird before picking up the second so as to avoid an escape of the first bird.

The problem that I have with people's understanding of animals is that we humans have an unfounded belief that animals are stupid when all signs point to them having a fairly high amount of intellect. The part that is even more perplexing is that those who believe in evolution should be thinking opposite of this and they should be assuming that animals could or even should have intellect close to or equal to humans, since the evolution belief is that humans ultimately evolved from some type of animal.

The fact that anyone assumes that animals are dumb is ridiculous. What is worse is listening to people evaluate human children in this same way, as if children are without intellect. Human children, even when they are babies, are without question brilliant far beyond animal creatures, and in some cases far beyond some adults, sad to say. Children learn at a pace unparalleled in the world, far faster than do adults. And animals tend to be limited in their learning capacity while still doing their animal thing with precision. Just think of how brilliantly a bird navigates the air in winds, or how the fish navigate the waters in strong currents. Animals do what animals do and they do it well. Animals are very intelligent in their own right, but to call them stupid or dumb or imagine that they are so by being surprised at their ability to reason shows more ignorance on our part than theirs. Darwin finally realized that high animal intellect exists and was amazed by it only because he did not believe that animals had intellect until that point of his revelation. If he misunderstood animals up to that point, then isn't it possible that he misunderstood a great deal more about all of his observations? That possibility looms large.

Evolutionism is Brainwashing

We often hear the talking heads of pop-science promoting the virtues of evolution and indicating creation as "brainwashing children as they are indoctrinated with tales of creation." But let's slip that mythical glass slipper onto the other foot for a moment, shall we.

"Brainwashing" as defined by Merriam-Webster dictionary is defined as: "a forcible indoctrination to induce someone to give up basic political, social, or religious beliefs and attitudes and to accept contrasting regimented ideas." This is the absolute perfect description of what occurs in the scholastic realm regarding evolution as it tries to eradicate any talk of creation. All thought and argument pertaining to creation is suppressed and utterly rejected which is supported by government, and then the students supporting creation are verbally beat down and unjustly mocked ad hominem for even trying to remotely defend *any* creation theory. At this point the virtues of evolution along with all of its imaginative theories are then forced upon the now morally defeated students' freshly emptied minds. If that doesn't scream of brainwashing, then nothing does.

Evolution as taught to students at the dawn of the twenty-first century is brainwashing perfected and is elevated to a forced state religion. Darwinian evolution *is* brainwashing.

Darwin's Evolution Superstitions

Superstition is typically associated with magic or sorcery type beliefs. Evolutionists often try to force religion into that general type belief-set, deeming it "superstitious" as if the Creator is some sort of lucky charm that grants wishes like some Arabian genie. But is Darwinism any better than that unfair assessment of "religion"?

Here we get into what religion is. Is the Bible religion in that superstitious way? Is being a supporter of creation superstitious in that way? Depending upon how you define religion and prayer you could loosely classify belief in God as "superstitious". But if someone seeks an answer from God and thus looks at the Bible to find that answer, is that any different than pondering evolution and then seeking answers from Darwin's or any other such book? No, obviously not. Those two examples are absolutely equal, with the exception that the Bible being very old and having been attacked and stood the test of time for thousands of years, gives it considerably more credibility than Darwin's works that are full of questionable conclusions that even Darwin himself was unsure of. Further, there are millions of people who would testify under oath that their prayer to God had been answered. The subject of prayer is explored in *Understanding Prayer – Why Our Prayers Don't Work - The Prayer How-To Manual.*

While Darwin is not a supernatural being for which to be superstitious about, the followers of Darwin are far more vulnerable in this regard because their blind faith belief mechanics are identical to that of creation beliefs, and Darwin never claimed to be a supernatural being. This would make him irrelevant in that case.

Darwin wallowed in his own man-made evolution superstitions about nature. Right or wrong, he invented imaginings about nature that the key part of is only speculation alone and nothing more. Darwin himself might be flattered that so many people have been fawning over his ideas since his work was released. However, after reading his works the second time, I dare say that he would be incredulous regarding the blind-faithed nature of his followers. These Darwinian followers fail to critically analyze new data, and instead they force it into his original model that he himself offered with some amount of reserve. Darwin sought evidence; his followers breach that idea and seek only his word.

Darwinism is Foul and Dishonest

Darwinism is said to be an "elegant explanation" of how things came to be, but it is only elegant if you like hocus-pocus magic. Darwinism, or evolution as it is affectionately referred to by its adherents, has many flaws that are suppressed from the view of the delicate virgin eye of his followers. This dishonest tactic of suppression in schools and science circles is foul and is a heinous crime now being perpetrated by governments world round. Here again, I must reiterate that to consider Darwin's research and analyze it as the theory that it is, is not a problem, but the religion of Darwinism with its suppression of opposing thought *is* a problem. It might be hard to believe that in any free country you would hear such words, especially in America, but there are people out there suggesting that people who oppose Darwinism and its evolution theory should be jailed and barred from teaching. Does it not strike you as dishonest that a dissenting opinion would be suppressed in such a way?

When you see such dishonest tactics happen you know that something is not quite right in the ideology of the religion of Darwinism. Steer clear of this sort of religion and investigate the evolution-versus-creation topic on your own using your own actual logic.

Will Unbelievers Suffer Hell?

After reading Darwin's works it became clear that he struggled with his religion. Few people know this, but Darwin was not just some guy looking into evolution. He had studied to be a preacher/pastor/missionary and was partly on a mission to teach native tribes abroad about God. A part of his observations were of these native people and that when these people were taught about the Bible and God it benefitted them greatly according to Darwin himself. During his voyages around the world he had the opportunity to observe all sorts of pristine environments not yet obscured by the hand of mankind. It is in these pristine environments that Darwin was able to make his articulate observations of nature and animals and then write of them to a world that was hungry for something new.

Darwin's religious training gave him somewhat of a unique insight as he observed the various creatures. But as mentioned in a previous section, he did not believe animals had intellect until he finally came upon the realization after witnessing it as he detailed in his account of the hunting dog. Darwin struggled with the idea that humans who he knew and loved who had rejected God would suffer the torments of hell. He had to cope with the possibility that they would find themselves in hell and he himself appears to have also rejected God due to that mental dichotomy.

The problem with this is that it doesn't matter what anyone thinks, because if hell exists then we will have to deal with our choices. If hell is just an imaginary tale, then it simply does not matter what we believe. But regardless of which case is true **we** do not get to decide if hell does or does not exists, rather, we only get to decide *whether or not we believe* hell does or does not exist. This makes Darwin's decision of possibly having rejected God a potentially very bad decision. Interestingly, if you become an unbeliever in Darwinism, then hell for you is close at hand because your hell will be hell on Earth when your peers attack you for violating the sanctity of the book of Darwin.

Darwin is Their God

This is not intended to be unfair to Darwin the man, but it might come across that way. In the same way that some Muslims follow Mohammed with improper actions, and some people follow Paul of the Bible rather than Jesus the Christ, and then go on to pervert the words and actions of those leaders, so too is the case with Darwinism. For many evolutionists, and yes there are exceptions, Darwin is their god, and they hang on every word of Darwin. And like most religious Bible thumpers and Quran thumpers, most Darwinians also have not read their holy book of Darwin which they thump so vigorously. When I make these sorts of statements I am not making light of it all, although it may seem that way. If you doubt this then take them to task on some of their thoughts and see how they respond. You will find out real quick just how religious most Darwinians are as you are bludgeoned with the book of Darwin as it is thumped on the heads of any opposition.

Darwinism is their religion and Darwin is their god. And unfortunately, their beliefs are supported by a great deal of government coercion and government funding. A religion that can only exist through suppression and coercion is not good for one's mind. It then becomes a cult.

Floundering Evolution

Single-source long-age Darwinian evolution is a floundering theory that has outlived its usefulness. There is a constant grasping to correct anomalies and errors that exist in this theory. It is the perfect lie and it is constantly being adjusted within itself to compensate for its errors. Where else can you do this but in the scientific origins theories. As evolution errors come to light, adjustments are made in assessment and ages and layers, and all sorts of explanations are created to hide the truth, which is that there simply is not enough information to conclude that single-source long-age Darwinian evolution is accurate or absolute.

This type of evolution is now and always has been floundering. At this point it must be repeated that we are speaking about the very specific model of evolution that is used in pop-science. One of the big problems with evolution versus creation is definition of terms. What is "evolution"? If you ask that question to several people you are likely to get several similar but distinctly different answers.

Most simply put, "evolution" is little more than change over generations or change over time. But in the commonly accepted definition it is changes over time going from simple life forms to more complex life forms. Both of these are generally reasonable views that even creation supporters will not disagree with. The disagreement comes in with "scope". Scope of change is the amount of change that any type organism/creature can incur through any amount of generations. The question here with scope is, is there a limit to the amount of change that can occur within a species, kind, or other classification distinction. Meaning, are the various evolution charts accurate throughout?

Evolution is a Belief System

Evolution as typically accepted, which is to say life having evolving from a simple single cell to complex life is a belief system that far outstrips any belief of creation theory from the Bible. What is interesting about evolution is that it mirrors the reality of the formation of a single instance of animal life, but it stretches that gestation out over hundreds of millions of years and credits it to generation upon generation of life. Could evolution as described be real? Possibly, but with the current rationale that it is derived from, namely suppress and oppress all descent, we will likely never know.

Ideas can only be proven when challenged thoroughly to screen for errors, but the errors found in evolution are rejected by Darwinians and are not allowed to be discussed or tested by opposition. There is great misunderstanding in the so-called scientific method in pop-science. A common phrase often heard early in the twenty-first century was "follow the science." By using that phrase they were able to hijack the term "science", and when you heard that phrase used, if you had any logic in you, you understood that it was a warning of vast scientific error perpetrated by those using the phrase. This would not be so frightening if you knew that they were deliberately lying about things. And some people were *specifically* lying. But what is truly frightening is that most people hearing this term actually believed that the science was being followed, when it clearly was not.

The kind of Darwinian evolution beliefs being pointed out in this book are damaging to young minds. People should be warned about this kind of Darwinian scientific fraud. They should be alerted that it is not settled fact and that much more study and research is yet needed to legitimately conclude the implications of Darwinian evolution.

Evolutionism Opinion versus Facts

If you actually take the time to read the book of Darwin that so much of the theory is based upon, you should quickly notice two prominent functions. The first is that of *observation* and the second is that of *conclusion*. These two, *observation* and *conclusion*, are not mutually exclusive. One is based in *fact* and the other is based entirely in *opinion*. Both are okay and are typically used on most any scientific path. In other words, we need to first observe and then make an assessment about the facts that the observation viewed.

Here is an example for you to consider: If you see a bird that is colored red, then you can make factual statements that you saw a bird and it was red in color. You were certain that it was a bird because it flew away and had left a feather behind. You can also state the fact that you saw this bird as it landed on and flew from a red flower. All of the above are factual statements about your encounter. However, if you now start to think about how that bird became red and obtained its wings and feathers and you then begin to associate the red color of the bird with the red flower and that somehow that bird was able to eventually fly due to generations of evolution of the wings and became red due to the evolution of its color, then that is *speculation* and your speculative opinions could be entirely wrong.

Pity Evolutionists

If you are at all a compassionate human being you have to pity Darwinians because they do not know any better. For many of them, if not most of them, it is the only thing that they have ever been allowed to be taught or have ever heard, much like natives of a primitive tribe. They might have heard other seemingly similar evolution theories, but to them it is all simply "evolution", so it is unlikely that they would notice any differences. And as far as creation theories are concerned, for the most part, Darwinians are not allowed to hear those theories, and if they are they are fed only those theories that are the most obviously inaccurate and they are immediately told that those Biblical based theories are fantasy. Have pity on these poor souls and be kind and gentle with them and teach them that evolution is not written in stone as they were taught, and that evolution has more than one perspective or theory. If Darwinians are able to have such seeds of truth planted in their minds, then those seeds of truth will most likely begin to grow and at some point a person with those growing seeds will have their eyes opened to truth. This will allow them to see the errors that were force-fed to them for the greater part of their life.

Darwin's Task List

When you dissect Darwin's work you will find a distinct pattern of observation action. Following is his pattern:

- Observation
- Immediate analysis
- Prediction
- Retro-prediction
- Extrapolation over long ages
- Extrapolation of effects
- Assumptions
- Final Conclusions

It is these mental actions that have allowed evolution to get out of control in Darwin's own mind as well as in the minds of many people in the general populous. And as a theory set of actions these are generally okay up to the *assumptions* point. It is at the point of assumption where his train goes off the rails. And when his train hit the wall of *conclusion* it became the wreck Darwin affectionately called "*Origin of Species*".

Chapter 22

Choosing Partners

Natural selection is partly, but largely, based on mating habits of creatures where the creatures will select what are the stronger or more attractive mates, which makes a lot of sense. It also makes sense for the purity of the propagation of a Biblical "kind". If there is a Creator who commanded things to reproduce "after their kind" then we would expect nothing less than natural selection of the most attractive specimens of any one "kind". The hijacking of this Biblical creation function by Darwin is theft in the highest degree.

Darwin noticed that the creatures he observed would often be vying for the attention of the more attractive, healthier, stronger mate and that mate would seek the more attractive provocateur. This principle is in perfect alignment with stable kind/species propagation that is fully consistent with the Bible's creation account, as well as with what is seen in nature. If any species seeks to the best of its ability to find the healthiest and most attractive mate, then that species is unlikely to ever breach the scope of the definition of that particular species of creature. And in that case a robin will never be anything but a robin.

No Really, Trust Me

Darwin did a great deal of observing on his voyages, and a hungry public was enthralled by his reports. As he traveled the world he would report back, to an anxious society, the many tales of his voyages. People trusted his observations because as he put it, "there can be little doubt".

All science must make suppositions when proposing theories, so it is more the problem of the adherents than a problem of the theorist that invents dirty scientific waters. But let us take specific note of Darwin's own language used in his own writings. His books are filled with concluding phrases such as *"If this... then there can be little doubt"*, or *"When this... we may conclude"*, or *"If this view be admitted there can be little doubt"*, or *"If so, then probably"*, or *"In all probability"*, or *"It appeared to me and I do not doubt"*, and even boldly stating *"Whether or not the half progenitors of man..."*, which places his assumption in people's minds that these *"half progenitors"* are a factual find, as does *"If so, before the progenitor of man..."* All of these seemingly "eloquent" statements are *not* statements of fact. These and many more such statements are statements of Darwin's conclusive assumptions that are merely speculations of a single person with no authoritative documentation whatsoever to back up those assumptions and conclusions.

He was in essence supposing people would accept his blind suppositions. And accept them they did. Darwin was the pop-scientist of his time and he was very popular with his followers of that time as he is today.

The Rotting Tree of Life

I once heard someone say that "the tree of life is now verified as fact by our decoding DNA". This is easily as inaccurate a statement as many of Darwin's assumptive statements were. Correlation by DNA association is not proof of descent; it is however proof of common attributes of kind/species and of fundamental function. I have heard people say that we share DNA with grass; Then does this mean that we evolved from grass? To some evolutionists, yes, it does. I have read estimates that humans have of up to twenty-five percent similarities in DNA as grass. Huh... even that lends to the Biblical man is made from dust of the Earth hypothesis derived from the Genesis One creation account.

But keeping things a bit more stable here, let us again visit the idea of shared systems. If you closely examine the structure of a brick and find identical bricks used in a house as you find used in a skyscraper, then does that make the two buildings related in any way? The obvious answer is, no, they are not related. But they do share structural components that are identical. They use these premanufactured functional components as a part of their makeup because those components work perfectly for the purpose of creating such structures. Using evolutionary logic, you can reason that you are identical to grass because both grass and humans are carbon lifeforms. And why are they both carbon life forms? Because carbon is the building block of choice and it works! This is a principle that most engineers understand very well. You design with the materials that you have available to you and which will work for the functions that your blueprints require.

A child's toy car that contains steel shares "DNA" with a skyscraper because they both contain the element iron in their "DNA" blueprint. All common building blocks do not demand linear generational association but are undeniably common and are used amongst many creatures in the animal kingdom. This is true both Biblically and evolutionarily and undeniably so. Functional design components are shared amongst creatures and amongst plants. Not only are elements shared, but DNA strings/functions are shared as well. Shared DNA strings do not make two creatures related any more than shared atomic elements do.

It is vastly speculative that the various evolution charts are accurate as presented, this does not say that they could not be accurate as presented, but rather that it is only speculation in the exact same way that Darwin's many presumptions mentioned in the last section were full of supposition and speculation. Just because systems share properties does not signify linear generational correlation. We see shared functional parts in physics, so why must we insist that it is any different in living creatures? Component parts share aspects of their blueprints or DNA just as one star shares the properties of another star simply because they are made of atoms.

The trees of evolutionary life as presented have rotten members in their branches. Many of these rotten members need to be pruned out of these trees of life before the trees die off completely.

Cladograms

When studying the trees of life you will encounter "cladograms" and "phylogenic trees" amongst others, these two are both trees of evolution, but they do differ in their basic functionality. The cladogram is the simpler more understandable diagram showing the names and speculative lineage of a given clade or animal group. Where the phylogenic tree is a bit more involved and gets a bit deeper into the inner workings, habits, and morphology of creatures. Both are interesting and both are speculative.

Undeniable Connections

Creatures are undeniably connected to every living ancestor of their kind in an unbroken chain as is proposed in both evolution and in the Genesis creation account, but that is all we know in "absolute" terms. We do not "know" that we are connected to monkeys or that chickens are related to dinosaurs as is often claimed. All creatures made from component parts, such as DNA and carbon, share these attributes. These common component parts that make up all organic material are common throughout creation and do not automatically constitute any direct linear descendancy. They show strong indication of component design and shared usage undeniably consistent with the Biblical creation account. But again, this does not "prove" that the Biblical creation account is what actually occurred, it proves that these common components and traits are *consistent with* Biblical creation and that is all that they prove.

Once a Biblical creation supporter steps outside of this basic information that is clearly stated in the Bible, they then become as Darwin was with his suppositions, assumptions, and potentially erred conclusions. Promote theories as the theories that they are, but keep them separated and far from actual facts.

Chapter 23

Why Didn't It Happen Some Other Way?

When we look into the origins of life, we see commonalties and we make certain assumptions about the commonalties we formed in our minds that are dependent upon our personal worldview. It is that worldview that trips us up. Upon examining findings surrounding origins we see many shared traits, functions, and building blocks as described in the previous chapter. We look for cause and reason and often wonder "why didn't it happen some other way?" But in wondering why it didn't happen some other way, we catch ourselves in the *why* versus *how* trap. *How* didn't it happen some other way is almost a nonsensical question, because there could be countless causes for it to not happen some other way regardless of whether it was creation or evolution. The question, why didn't it happen some other way, seeks reason and causal purpose.

We humans have an unquenchable desire to understand our origins, and we make our attempts at that understanding through evolution or through Biblical creation and we have been doing so for centuries if not millennia. In seeking our answers of the causes of origins, we in that desire alone should take note of our quest. There is a great deal of information to be had for those seeking it when we analyze our own motives regarding the question of evolution or creation.

To answer the "why didn't it happen some other way?" question, it is because very articulate systems were in some manner established and those uniquely established systems allowed for certain functions to occur. Those functions led to the ability for reproducible cells to establish themselves in such a way that would allow for life to be and propagate. These cells required some sort of instruction set to follow in order to consistently reproduce themselves and form themselves into the larger forms of the various creature types listed, which is consistent with both Biblical creation and with evolution.

What Are Eukaryotes

A eukaryote is cell or group of cells that have genetic DNA in them that form chromosomes in their nucleus. Nearly all types of living organisms are eukaryotes. This type of reproducible organism is consistent with both creation and evolution. But in the case of such cells, they are capable of reproducing themselves through cellular division as when a female egg is fertilized by a male sperm. Once the egg is completed by the sperm in this way it will continuously increase its quantity by dividing as it extracts the required building blocks from its surrounding environment and the instructions from the contributing cell. These cells themselves are monophyletic; however, they typically are the result of a polyphyletic pair in humans.

Law or Principle of Monophyly?

I once heard someone refer to Monophyly as a law sometimes referred to as Darwin's law of common ancestry. Let's investigate this thought. This law or principle basically states that one branch of life or species cannot give birth to a different basic kind. This key tenet of evolution is in absolute agreement with Biblical creation in that creatures give birth to offspring creatures of their own kind and do not spontaneously give birth to another kind. Since ancestry is an absolute unbroken chain of succession it is not possible for that ancestry to ever change, but this does not mean that the potential for *speciation* through subsequent generations is not possible.

The immediate problem with the idea of *speciation* is again that of definition. What is the specific definition for a "species" or a "kind"? "Speciation" is the theoretical change in the form of a creature over generations of change that raises cause for a new species distinction to be made by us humans. Where this new species line is drawn is a somewhat of an arbitrary art, though that would be argued by most evolutionists.

Here is the problem with speciation and species definitions or distinctions: What is the level if gradation? How fine or granular of a distinction will we establish between species? You can picture this in terms of the increments on a meter or yard stick. A yard stick and meter are pretty close in length, but a yard stick will often display only 1/8-inch increments leaving a total of 288 marks on the yardstick. A meter is sometimes presented in millimeters, giving us 1000 marks in a similar distance. Or you could even use meter sticks, having one with millimeter marks and one with only decimeter marks that is to say 1000 versus 10 marks for the full meter length. One of the meter sticks has 1000 divisions and the other has only 10. Regarding evolution, who gets to decide the gap of distance or difference between species designations with each species being a line on the species rulers? Will we measure species relative to the millimeter example, or will the species be determined relative to the decimeter example?

There are some compelling and also some undeniable species distinctions, but most of those species distinctions are ultimately a matter of distinction opinion made by differing people with differing opinions on differing species meter sticks.

The True Definition of "kind"

Because the speciation issue is one of the primary stumbling blocks that cause the great chasm between evolution and creation, we are again going to try to arrive at reasonable and fair assessment as to what such words mean.

The term "kind" has an etymology that implies "kin" or "family". The obvious implication of the usage of "kind" in some translations of Genesis is that these "kinds" are related or are a family. This is fully consistent with both creation and with evolution. Then we have the Latin term "species" here is the translation text for your review:

English Douay Genesis 1:25

"And God made the beasts of the earth according to their **kinds**, and cattle, and every thing that creepeth on the earth after its **kind**. And God saw that it was good."

Latin Vulgate Genesis 1:25

"Et fecit Deus bestias terrae juxta **species** suas, et jumenta, et omne reptile terrae in **genere** suo. Et vidit Deus quod esset bonum.

"Species" has an etymology that suggests examination of the way something looks, meaning to *spy* or *see*. So, both of these terms imply clearly that the creatures are similar to their particular grouping concerning the Bible's creation account, meaning that the offspring will nearly identically resemble the parent creatures. Where we immediately run into trouble is in the gradation or increments on our species-ruler example. How many kinds are there listed in the Genesis One creation account? Here again is the actual Douay English version listing of the kinds for your perusal:

- creeping creatures
- winged fowl that may fly
- living and moving creatures
- great creatures (whale)
- living creature

- cattle (Greek – tetrapous, Latin - jumenta)
- creeping things
- beasts of the earth

Any supporter of creation, or supporter of evolution for that matter, that steps outside of this basic Biblical creation definition framework is writing things into the Genesis One text that the text simply does not say. This does not mean that finer resolution "kinds" are not included in that listing, such as pigeons, seagulls, and hawks or kites etc. While that possibility does exist, it is simply not included in the Genesis text, nor is any deviation from the very few particular kinds that are actually listed or described therein.

When we get into the scientific evolution description, the resolution tends to get increasingly higher as the years pass. So the Bible is measuring in decimeters, where evolution on the other hand is measuring in milliliters or more, that is to say a few choices or divisions per species meter, versus thousands of choices or divisions per species meter.

Hung Up On Scientific Terms

I remember seeing a sign once that said "If you can't dazzle them with brilliance, then baffle them with BS". Not meant as a dig to evolution, but this is basically what often occurs in debates of evolution versus creation. Ardent evolution supporters have an affinity for evolution *jargon* and will spew that jargon at their debate opponent as if their mouth was Niagara Falls. It's good that they know these terms *if* the truly understand them, but often these terms are irrelevant in the course of the debate. People get hung up on these terms, many of whom have paid colleges tens and even hundreds of thousands of dollars to teach them such terms. Refunds might be due for many people for colleges selling faulty goods.

So many of the terms like *cladistics, mono-cladistics, phylogeny, monophyletic, polyphyletic, eukaryote,* and so many more are all basically smaller parts of the millimeter increments on the species meter sticks spoken of earlier, or they are a second meter stick ever more finely incremented for measuring a different aspect of the first meter stick. If we are not going to nail down terms such as "species", then what is the point of science? Science of the past was successful not because they invented an endless string of new terms, science of the past was successful because physics and biology are *consistent* and *reliable*. People evolving terms like is done with "species" is inherently problematic when trying to debate.

Regardless of the language or particular words used to translate the Bible, the fundamental fact that the kinds mentioned in authoritative Bibles do not change is the Bible's key to its longevity. Those biblical "kind" divisions, of which there are eight, at most, depending upon how you group those words, remain constant and they do not change and have been written for thousands of years. These words are consistent with evolution and creation and they do not veer from their purpose.

Chapter 24

Eternal Consistency

You might have read through this entire book and be disappointed that you did not get an absolute confirmation that creation or evolution occurred either this way or that way. I will offer a likely conclusion shortly; however, I will do so with great reserve in the *Final Thoughts to Consider* section.

Regardless of how you frame this debate, when using an older authoritative Genesis One account that has **not** been perverted by the more recent post-Reformation Bible translations, unlike evolution theory, the Bible, right or wrong, has remained consistent. However, people's interpretation varies wildly depending upon their upbringing, schooling, and their preferred Bible version. But do not let other people's poor interpretation sway you from trying to understand the Bible's creation account. Much of the misinformation about the Bible is from evolution supporters who insist that *their* interpretation of the Bible that they grew up with is what the Bible says. I have found those interpretations to be typically untrue.

When you fluster people in debate they begin to flounder and grasp at straws to stay afloat. Evolutionists will begin attacking God for no particular reason, and all too often the foolish creation supporter will take the bait and begin to defend the God that they claim is "all powerful". A true God or Creator needs no defense, but our theories often do. When a creation supporter gets flustered they too will begin to attack from illogical perspectives that have nothing to do with the subject at hand. Steer clear of these ad hominin attacks and learn to detect them for the sake of scientific progress. Scientific progress in modern times is typically credited to evolution, but in truth, the most significant advances in scientific progress over the years came from creation supporters of past ages who were dedicated to their God and to the creation account. They sought to better understand their Creator. Part of that understanding, sad to say, was the rewriting of the Genesis One creation account in the post-Reformation Bibles mentioned throughout this book. This misdirected zeal has created a hostile group of Bible deniers that behave as if they are a part of an evolution cult that is ever accusing and never listening. This is fits well with Isaiah

From Douay English Isaiah 6

"[8] And I heard the voice of the Lord, saying: Whom shall I send? and who shall go for us? And I said: Lo, here am I, send me. [9] And he said: Go, and thou shalt say to this people: Hearing, hear, and understand not: and see the vision, and know it not. [10] Blind the heart of this people, and make their ears heavy, and shut their eyes: lest they see with their eyes, and hear with their ears, and understand with their heart, and be converted and I heal them."

The most prominent problem on both sides of this debate, especially on the evolution side, is that of refusing to hear. When people are forced to hear the words and begin to see the light of truth as it begins to penetrate their coats of evolutionary armor that they have forged for themselves, it results in those ad hominin attacks on God just mentioned that do nothing to further the debate. Unjust mockery makes a fool of the mocker.

Impotent or Omnipotent

When frustration ensues and the unjust attacks begin, we hear things like "god is impotent because over nine-million children die every year from disease and hunger." Such statements are at best illogical and at worst outright ignorant. But this is understandable since most people who say these sorts of things also greatly misinterpret Genesis One. If we are going to speak about this God of the Bible as if we believe in God then we must adhere to the text of the Bible rather than grabbing inaccurate thoughts and combining them with our misguided opinions about the Bible in such an ad-hoc manner.

In reading the first few chapters of Genesis describing the brief account of creation and the arrival of mankind, you will quickly see that a critically important point was the freewill of mankind to make choices. God did interact with people in the Bible, but it was usually at arm's length to allow the people to self-adjust to decency. This Creator spoken of in the Bible did not build a fleet of robots designed and pre-programmed to return love on command. Humans were very specifically given free will to, or not to, return love to both God and to our fellow man. Trying to pin the blame on God for our human deficiencies that causes the deaths of millions of children every year is inherently dishonest. It is we humans and our fighting that cause this problem. Very few of the areas of extreme poverty in this world are environmentally deficient. Most disease and starvation problems are due to wars and the shortages and deficiencies in resources that those wars cause.

Further, if we are to blame God or consider God impotent due to the annual deaths of so many children, then we best also consider two other points of concern. The first is that if the Bible is true and there is a God/Creator then there is also a Heaven to which innocent souls would ultimately go. This issue of the final destination of souls is said in the Bible to be of more importance than the life we live here on Earth, so the deaths of these children that we adult humans alive today are responsible for is not a big concern to God because they will be "saved" and dwell in Heaven with God if they are innocent. The other point to consider for those who would make such an accusation about those poor starving children who die every year is that it is our fault that those children die, but that's not the worst of it. What about the other roughly fifty-million children who die every single year from murder in the womb under the guise of "abortion"? As a free will people, are we also going to place the blame for our sins of the murder of roughly *fifty-million children annually* on God? Will we blame God as we riot and protest in the streets to demand our "right" to murder our own offspring?

These sorts of unfounded attacks that commonly occur in the evolution versus creation debates are unfair and do nothing to further science. It doesn't matter who is doing the attacks, the attacks are a retreat from science rather than an advance thereof, and they get us nowhere and cause nothing but frustration.

Science is Retreating from Science

"Science" at one point in time indicated "knowledge", or at least the quest for knowledge. But what do we call something parading itself as "knowledge" when it is clearly wrong? In Hosea 4:6 it says "My people have been silent, because they had no knowledge: because thou hast rejected knowledge, I will reject thee". This indicates that we are to pursue knowledge, thus we are to pursue science. In the Bible the obvious intention for humans being created in the likeness of the Creator is **_to get to know the Creator_**, which was the goal of most of the early creationist scientists. The Bible versions that have been greatly distorted that I keep mentioning are a serious retreat from true science. They are a retreat from actual knowledge.

If we are discussing things from a Biblical perspective in this regard, then science has been retreating from God for some time now with all of its godless evolution-speak. In in a Biblical sense retreating from God is retreating from the ultimate scientist and is ultimately the rejection of God, of science, and of the knowledge spoken of in Hosea 4:6 just mentioned and that is so perfectly illustrated in Isaiah 6, here it is again for your convenience:

"[8] And I heard the voice of the Lord, saying: Whom shall I send? and who shall go for us? And I said: Lo, here am I, send me. [9] And he said: Go, and thou shalt say to this people: Hearing, hear, and understand not: and see the vision, and know it not. [10] Blind the heart of this people, and make their ears heavy, and shut their eyes: lest they see with their eyes, and hear with their ears, and understand with their heart, and be converted and I heal them."

Yes, those Bible verses sum up the entire evolution-versus-creation argument so very well. And, sad to say, it is little different no matter which side you happen to take in this debate, because both sides, as far as can be determined, "Hearing, hear, and understand not: and see the vision, and know it not", and in doing so they stunt their knowledge and inhibit their scientific advance in better understanding our actual origins.

The Shallow Science of Evolutionism

Darwinism or evolutionism is a shallow religion with an I-am-right-and-you-are-wrong overall philosophy. It is a petty religion of the unseeing and unknowing. There is no true science to evolutionism, which is to say no true knowledge to evolutionism. We cannot *know* something that is wrong. You can know that something is wrong, but you cannot technically know information that is wrong, because, technically, it does not exist. You can understand the wrongness and why something is wrong, but wrong things of evolution do not exist.

The self-aggrandizing practice of evolutionism is readily apparent nearly immediately upon its adherents opening their mouths with their fountains of folly pouring forth from their mouths with the rush of a torrent of their overpriced vomit taught to them by their "education". Evolutionism is shallow and wanting and demanding, but offers nothing in return.

Let us separate *evolution* from *evolutionism* and come to understand that change to some extent does occur, but that it might have limits at least to the extent of the very few kinds listed in Genesis One.

What *evolutionism* thinks the layers of geology tell us, versus what geology actually tells us are not in harmony. If you are interested in what the geology might be telling us take a look into it without belligerently denying other points of view, such as the views discussed in *The Science Of God Volume 5 – Boats, Floods, and Noah – The Deluge*. Consider it and other people's points of view and, most importantly, look at the evidence for yourself leaving nothing out, and then form your own opinion instead of automatically regurgitating other people's words. If you fully review their information and happen to agree with it, then repeating their words is not a problem, but you should at that point be able to form your own thoughts and sentences when you discuss the topic, at which point you will then have little need to repeat other people's words exactly as they say them, because you can describe it all in your own words.

We Come to a Point of Belief

If we study a subject and understand it properly, we should be able to dissect it and debate it with intelligible statements or at least concede a point when we are unsure by simply saying "I don't know." If you have a topic well understood you should be able to catch the errors of your debate opponent. However, if you know the jargon and know the meanings of the jargon, but ultimately do not really understand the subject, then you will most certainly be beat down by a worthy opponent. But even if you defeat your opponent but do not truly understand the data, you will likely only be able to offer old severely worn-out platitudes.

All of the nauseating pop-science evolutionism debates come down to a point of religious blind faith beliefs. Everything in life can, in a philosophical manner, be said to be a point of personal belief, but let's not step that far out of the topic. Most people have a blind faith belief in their chosen perspective on the evolution-versus-creation topic. But if you use your gift of basic human logic, you should be able to see certain undeniable truths emerge that apply to the topic. Some of these undeniable aspects of the Genesis One creation text were mentioned throughout this book. These undeniable points in the text are not so much saying that evolution does or does not occur, but rather they offer clear indication that Genesis One as stated in authoritative Bible versions is undeniably accurate. However, the perceived errors in it are due to our own poor interpretation attempts that we then forcibly cast upon our neighbors.

The facts you choose to extract from the Bible's creation account and the things you choose to believe are your own responsibility to accurately assess. If you choose wrongly and then proceed to teach your inaccurate view of the Genesis One creation account, or even evolution, to others, then you have made yourself a false teacher, and potentially a liar.

Points to Consider Rather than to Ignore

Probably the most damaging problem in all of this is that we ignore what we don't want to hear. There is a great deal of information to consider rather than ignore. Look at the geological layers that establish the order of events for evolutionism. Really study them and cross-compare them from other layers around the world, rather than assuming that evolutionism "science" is right. Some theories have enough holes in them to sink an entire ark.

Realize that Darwin's books are filled with suppositions such as, "*if this, then wouldn't that be possible?*" These arguments are persuasive because they move the focus from the real issue to a logically undeniable hypothetical issue. This tactic is like saying if someone gave you a billion dollars, then isn't it possible that you would be rich. You have little choice but to answer "yes" to that question. However, this deceitful practice ignores the unlikely case of someone just randomly giving you a billion dollars. The majority of Darwin's work is framed in that type of logic, and many unsuspecting people have fallen prey to it and have been lured into the church of Darwin whose religion is evolutionism.

Another point to not allow yourself to get caught up in is the erred understanding of what "evolution" is. People can talk of macro- and micro-evolution, but in the end evolution is nothing more than change over time, which is a debate that has generally always been aimed at **the amount of change** that occurred or that is allowed to occur as a result of the evolution.

Steer clear of using the Darwinian logic spoken of earlier with its "if this, then isn't it possible?" logic. Using that logic method is like saying if there is a God then wouldn't it be possible that God did [*this or that very illogical thing*]. Darwinian logic can gain you an upper hand in the immediate argument, and unsuspecting people might even fall of your deceit. But you have advanced nothing scientific about your theory except for forcing people to agree with such unscrupulous dishonesty.

Be cautious when in debate or discussion about the evolution-versus-creation dichotomy. Evolutionism is filled with tools to thwart your every word. A randomly occurring tactic is to change the debate through insertion of evolutionism and long-age geological evolutionism beliefs into the Bible. When doing this your opponent will project their erred theories into the Bible by basically insisting that you accept that their geological age estimates are perfect and correct and that the Bible does not match that data, therefore the Bible is wrong. This is most notably done in the Noah flood debates that have strong connections to the creation debates that are associated with the geological record of mountain altitudes, timeframes, continental movement, and so much more which are all addressed in *The Science Of God Volume 5 – Boats, Floods, and Noah – The Deluge.*

The mastery of deceit, whether deliberate or accidental, that people who thrive on evolutionism practice is really quite impressive. Their goal is in no way connected to truth, but rather is intended to dominate the debate or conversation to prove that they are *right* rather than *correct* and *accurate*. If they cannot accomplish their goal through their mastery of deceit, then they will typically turn to using those personal ad hominin attacks previously mentioned. If that fails, their next go-to is usually all-out mockery and trying to get a laugh at your expense. Beyond that, if it gets that far, they become belligerent as they shout at you and demand that things be their way. I wouldn't believe these things of a civilized society had I not repeatedly witnessed them firsthand with my own ears and eyes.

Never forget that all of the graphic or modeled presentations of every fossil found are only the speculation of what an artist imagined the creature to look like based upon what are typically finds of incomplete fossils–There are no exceptions to this simple truth. The artist could not have been alive when the fossil was cast at the claimed hundreds of millions of years ago, therefore *they did **not** see or witness the creatures.*

With the exception of "great creatures" fossils such as dinosaurs, many of the fossils found that are deemed to be hundreds of millions of years old are nearly identical to skeletons from animals that exist this very day. While this does not disprove evolution, it does prove that some "kinds" have not changed much over those suggested millions of years, which happens to be perfectly consistent with the Biblical creation account.

When it comes to the topic of evolution versus creation, it will always eventually come down to the initial point of origin. Was it a random lightning strike hitting chemical gases that formed amino acids that then somehow randomly began to generate and form all of life as we know it today? Or did the Creator specifically create, at minimum, the quantity of "kinds" listed in Genesis? We may never know for sure.

Final Thoughts to Consider

The truth of the matter is that our minds have been so dirtied with far-reaching prominent theories from either side of the evolution-versus-creation chasm that we now struggle to clear that fog from our thinking. If you do deep consideration of this subject and really think it all through and then were to re-read this book, I suspect that there are many thoughts relayed within it that would stand out to you much more prominently the second time through. Clearing the foggy pollution of indoctrination from our thinking is indeed a very difficult task for us humans to do on our own. It is my hope and prayer that those who read this will have fresh and pure seeds of true logic planted in them so that those seeds might grow to outstrip the lies that most of us have been indoctrinated with throughout life from both sides of this debate.

Is There An Answer?

Is there an answer as to whether or not life was created, versus having evolved by chance? Your answer to that all depends upon your take on life. I often hear creation supporters that are Christians saying things like "we can't use logic alone" and that "we need faith". There may be a bit of truth to that regarding the Heaven of angels and all associated with it, but when it comes to the tangible physical aspects of creation, logic can and does prevail if you are able to abandon blind faith. Blind faith is perhaps more prominent on the evolutionism side of the spectrum than on the creation side. Logic and geology do tell us a great deal about the advent of creature-life when we choose to make the effort to hear what it all has to say to us, rather than the typical act of us trying to tell logic and geology what to say. If you pay close attention to this topic you will quickly realize that most people who present their case about the evolution-versus-creation topic are *writing* the story rather than *reading* the story as told by logic and geology.

Are there answers? Yes, but we must zero in on the real questions. The primary question is, how did we get here? And also, does God exist? How can there be two questions that are primary?

If we deny God we are absolutely stuck in evolutionism because then there are no other choices of what path to follow. If we accept the concept that a Creator-God exists then the typical assumption is that the Creator magically made things instantly appear, POOF!, and it was! A magic hocus-pocus God that behaved like a magician is not logical in any way. Is this saying that it could not have occurred that way? No, not at all. However, given all that we see and the way that it all behaves, and given the Bible's text, and given our supposed human "in the likeness of God", it is highly unlikely that hocus-pocus occurred anywhere in creation. We do not witness it anywhere!

Yes, there were a few "miracles" done in the Bible, but those were rare and most can be attributed to the Creator's ability to interact with Creation. We simply do not know the mechanisms that may have been used by God in parting a sea, for instance. For us to insist that something was one way or another, in that particular case, is, again, writing things into the text that simply do not exist in the text.

After extensively examining the evolutionism side, and the Biblical creation side, and after considering the remainder of the Bible, I am left to conclude that this Creator that we call "God" is extremely *logical* and *fair*, and has a nature that is similar to us humans that we were instilled with during the "in our image" action, as further explored in *The Science Of God Volume 4 – Day Six – Evolution versus Man – In Our Image*.

Can we determine if God exists by exploring the evolution versus creation topic? That is truly going to depend upon what you are willing to accept as evidence. I can point to the ground you walk on and say that it is evidence of the Bible's creation account and therefore of a Creator, which is actually a fair point to be made. However, that particular point will be found as "irrational" when discussing this with someone who practices evolutionism.

Much of this comes down to blind faith. Many people might argue this point, but it also takes blind faith to *not* believe in a Creator. Like in the book *The Science Of God Volume 1 – The First Four Days* I like to take people back, way back, to the initial point of creation or "big bang". Even though in big bang theology the point of "singularity" is said to have been so small that it would essentially have been undetectable, big bang says from that singularity point of nothingness a "big bang" occurred and eventually produced all things. From a debate perspective I can accept that, even though I do not agree with it. But at the moment before, or at any time before this supposed big bang occurred, how did the point of "singularity" get there? If you say it was just there or that it would have been undetectable, then you have blind faith, there is no other logical conclusion about that position. Godless singularity is blind faith on steroids and it far outstrips the logic and faith required to accept the idea of *Eternal Mind* that created all.

The Genesis account taken from authoritative Bible versions is scientifically logical in its order of events and its allowable-scope of description, but it faces the same logic issue just mentioned regarding the singularity of big bang theology. That is to say, how did the first elements come to be whether from singularity or creation? We mentally oscillate on this. I have seen it many times where people struggle because they logically understand that something, no matter how small and insignificant, must have somehow been placed there. But if placed there then by what or by whom, and if something or someone did place it there then where did that come from, and again if something placed that there then where did it come from, and on and on it goes for infinity. This leaves us in our human logic with only one logical explanation and it is not a big bang.

Evolutionism seeks to reduce complexity in effort to make it believable in the minds of its followers. A part of evolutionism is big bang theology. These two inseparable religions of big bang and evolutionism take the entirety of the universe and compress it down to a point of insignificance within its adherents' own minds. And then because of its insignificance there is no need to any longer consider it of relevance or explain it, thus the irrelevant singularity can then explode with the force of big bang to create a universe in which evolutionism can then proceed. Then the same thing is done with evolutionism in that it takes some seemingly insignificant chemicals that arose out of an insignificant singularity and from those insignificant chemicals lightning then ultimately creates life.

Here is where this gets interesting: Some of this insignificant material and activity could have been used, even potentially in the way that is promoted by evolutionism, but could have been done so by a Creator. This is where much caution must be used when analyzing the abundant, yet limited, data that we all have available to us. If you are being logically honest, you will not wonder if a Creator exists, but rather you will wonder how the Creator did it. It is the *how* that we are all trying to drill down to regardless of the existence of a *who*. However, we instead get tripped up in this senseless debate of whether or not a Creator exists. And regarding matter, it is logically obvious when simply repeatedly asking "how did it get there?", which brings us to the creatures.

How Did Creatures Get Here?

How did the creatures get there? Could it have been lightning randomly hitting chemical gases resulting in amino acids thus allowing the life process to begin? You bet it could have happened that way... but did it? We may never know about those initial moments, still there are more important questions to answer about the arrival of creatures. The lightning striking chemicals theory is interesting and worthy of exploring, but the way in which it is promoted is left wanting.

Here are the questions that we must answer, of which I personally believe that the answers to these questions are somewhat logical.

First, whether through lightning striking chemical gases or any other remotely similar means, the key question here is, was this an isolated occurrence or did it occur abundantly everywhere on Earth, potentially including on other planets in other solar systems in this vast universe in which we live and explore?

Next we have to answer the question of *scope*. Regardless of how it began, did life start from single sources, meaning did "kinds" arise from a single pair or were there many similar instances of similar coupling of creatures? From a creation perspective this is asking if the birds that were brought forth were only two (one male and one female) and all of that "kind" came from them, or where many of each gender, both male and female, brought forth right from the start of each kind?

The next question that we must ask is that of time. We are not so concerned with millions of years versus billions of years here but rather, did whatever occurred do so instantly, or did it take time? This question has two aspects; the first is, were these creatures fully formed immediately? And the second aspect is, if not, then did they have to morph into their "kind"? Here is the summary of the questions we must ask, the three questions are:

1) Did life begin at one location or many?
2) Did life begin with many of any one kind or species?
3) Did life occur quickly or was it slow?
   a) Did life initially arrive fully formed?
   b) If not, then did Life have to morph into a particular kind or species?

To answer question one you will need to look at the vulnerability of a single cell and reason through the likelihood of a lonely single primitive cell self-created of pure chance actually surviving. That simple mental exercise gives us a clue that life arising was probably *not* a single instance, but rather, that the process was likely occurring all over the Earth at that time in the history of our Earth as well as other celestial bodies. Our life experience confirms that life is robust and occurs *everywhere*, even if we try to stop it with insecticides etc.

As for the second question of life coming from one kind or from many sets of one kind, the logic is a bit more complex here because it is closely related to the third question's details. However, in trying to keep this simple and very straightforward, what we are ultimately asking is going to be based on the first question. In evolutionism it all began from a single-source, forcing the second question's answer for those believers to be "one"–it came from a single-source. Now, if the more logical answer for question one is accepted where life is robust and the initialization likely occurred in many places at the same time, then we are free to consider that it may also have produced a variety of initial kinds or species. The small selection of "kinds" mentioned in Genesis is certainly a reasonable selection quantity that does not force anyone to stretch their imagination.

On question three, evolutionism forces people to believe that it all took a very long time, and that expanded time frame is required for the mathematical chances to be possible for life to occur as is implied in evolutionism. On the Creation side of the debate it could have been a long time, but such extended time periods are not required as are used in evolutionism. Six-twenty-four-hour-day creation is quite unlikely in nearly every aspect, so we will not even address it here. But this particular aspect of creation could have been very long or relatively short, but it was very likely *not* instant. Life arriving quickly or slowly is a matter of relative definition of the duration terms *quick* and *slow*. In our case here, question three had two aspects, the first aspect asks if life came fully formed, and the second asks if life had to morph into its kind or species.

Evolutionism demands that it was from a single-source that slowly morphed generation upon generation virtually unnoticeable between any two generations, and then over millions or billions of years it slowly multiplied and formed the distinct species that we have today.

In answering these questions, creation allows for life to have begun in many locations around the world at the same general time period, which further allows for varying kinds to arise simultaneously from the start even if they had to evolve to some extent to produce the variances within each listed Biblical kind. And finally, the Biblical creation account allows for life to have taken a great deal of time, or it could have been relatively quick depending upon the answers to the two sub-questions under question three.

To answer question three in trying to assess whether or not life arose relatively quickly, let's first look at question three's part-b asking, "did life have to morph?" This is asking if life had to evolve from a primitive form into a complex form over many generations. Evolutionism would have you believe that it is scientific "fact" that life began billions of years ago and was required to gradually morph over time into the various creatures of today. This is possible even in the Biblical creation account, but is it likely? Because we are splitting hairs on this critical area in the overall debate, we are allowing all options in discussing what the Bible's creation actually says. The distinction of "kinds" or "species" is truly that of a "kind's" offspring, this allows for two logical options: one goes down the evolutionary path where each subsequent generation looks like the parent generation but with enough generations can morph into something noticeably different, and the other is the immediate arising of specific kinds. Both of those are possible within the creation account, but are both logical?

This is where evolution, creation, geology, and floods collide in violent manner. The geological record shows one thing, but evolutionism demands another. In evolutionism it is implied that there are tidy layers of sediment that over millions of years have slowly been deposited covering up and trapping all sorts of creatures in the sediment, thus the lower layers, being deposited earlier, trapped the most primitive creatures. Then as the creatures' complexity slowly evolved over time, the evolutionary variations of creatures would also become trapped in subsequent deposited sediment above the earlier layers. Then over millions and millions of years of this occurring, we now have a perfect and tidy geological record of evolution. This sounds great, except that things are not as tidy as is presented in textbooks distributed by the church of evolutionism.

There certainly are layers that are undeniably deposited sediment. And it is undeniable that there are creatures trapped within those layers of sediment. And it was an undeniably long time ago. But not all of the layers throughout the world agree with the church of evolutionism's geological column textbook models. Even one single compromise of a more complex lifeform below any primitive creature found in the lower layers of the geological column will completely forfeit the credibility of the entire premise of evolutionism. In reading *The Science Of God Volume 5 – Boats, Floods, and Noah – The Deluge*, you will find that this is why evolutionism so fears the concept of a global flood. If you ever want to test this, simply bring up the idea of a recent geological global flood to an ardent evolutionist and you will find that you will typically be met with a considerable amount of hostility and or mockery.

The explosion of life found in the "fossil record" defies evolutionism and more accurately illustrates that life sprang up fast, really fast, as modern geological models go. But since the topic of a global flood is often frowned upon I like to revert to logic in this particular part of the evolution-versus-creation topic rather than use flood arguments. So, let's forget about the flood and its implications for now, you can explore more about that in *The Science Of God Volume 5 – Boats, Floods, and Noah – The Deluge*.

From a Biblical creation account perspective, we simply *cannot* insist that at least some level of evolution did not occur, so we must define what that implies. Looking back at the idea of life springing up in many places around the world at the same general time, one could say that each instance of life that sprang up produced a different species, but that stretches the Genesis text far beyond its scope. When you really think about it, it's very logical and rational to consider that life may have sprung up in many and diverse places around the world during that period, which does not defy or write anything into the genesis account.

However, what we want to try to establish is, was that life primitive, or did life arise in its gestation in full form? Here we see two possible paths and both pertain to question three part-a and part-b. Life either morphed from primitive to complex, or it did not, there are no other choices. Morphing in a sort of sideways manner to get differing types of a particular creature like the various but very similar hummingbirds is still able to place those variations within the Biblical creation "kinds". We know for sure that life morphs or evolves, but we do not know for sure if there is a limited scope of the morph-ability of "kinds". I have heard on more than one occasion where practitioners of evolutionism angrily demanded a definition for a "kind" from creation supporters. To that I would say read Genesis One, but for convenience sake, here the kinds are:

- creeping creatures
- winged fowl that may fly
- living and moving creatures
- great creatures (whale)
- living creature
- cattle (Greek – tetrapous (4), Latin - jumenta)
- creeping things
- beasts of the earth

Any person who supports the creation side of the debate and who is honest will not step beyond these listed "kinds", and the same is true of anyone who practices evolutionism if they want to be honest about what the **Bible *actually*** says. While the particular words used in the various Biblical creation translations may vary somewhat, these general "kind" groupings are quite evident. That list is what the Bible's creation account shows, and anyone demanding more or adding various divisions into the listed kinds found in the Bible's creation account is a liar. Yes, harsh, but true!

A brief side-note that I found interesting: I backed up to this section while writing to insert this because all of the words written after this point were done on the final day of writing this book. No big deal, but I found it intriguing that when I went to verify something about creatures, the picture of the day on the search engine homepage was of a fossil from an Ichthyosaurs. That was an interesting coincidence, but what was more interesting was that the National Park Service calls that day "National Fossil Day" which I did not know. Not a big deal, it just was peculiar to be wrapping this this book up on that very subject on National Fossil Day. Now back to it!

The Bringing Forth

On day five the waters brought forth, and then on day six the earth brought forth, leaving us faced with trying to understand what was meant by "bring forth". In Latin, the word used is "producant" which is like our English word *produce*. Greek uses a term that would be pronounced "exago" meaning to *lead forth, bring forth,* or *fetch out.* Translation can certainly be a tricky task to tackle, but it is fair to infer from these languages that the "waters" is the vehicle used from which the day five creatures were derived or brought forth. If we are correct in that we assume that, geologically speaking, Earth's continents did move, then we can be certain from our current understanding of weather patterns seen today that are affected by landmass position, that weather patterns in those early days of Earth would have been very different than we experience today. Continental movement is very obvious when you study any imaging of the ocean floor that we today have the privilege of studying due to all of our technology. Anyone can study the ocean floor and the evidence of continental movement at their leisure using nearly any modern computing device. There is more on this subject in the book *The Science Of God Volume 5 – Boats, Floods, and Noah – The Deluge*.

All of what has been offered so far is logical and reasonable and all falls within the Bible's creation text and even within the greater part of the evolution theories without ignoring or stretching any of the Genesis One creation text beyond its scope. This next thought, while still remaining within sound logic and reason, does rely on words or understanding of the concept most likely intended in the text of any of the known ancient Genesis One versions. All of them seem to clearly indicate a bringing forth of or producing of life. We can equate this particular action with giving birth to or bringing life to creatures. The text clearly credits the "waters" for giving rise to the day five creatures. Latin uses "aquae" and Greek uses what is pronounced as "hydor", both of which indicate what we all know and understand as "water". The Greek can also be understood as *fountain* or some sort of pouring forth of water, kind of like a fire *hydr*-ant delivers water.

This idea of water bringing forth or giving birth to something is scientifically critically important to grasp and it is the tip of the fulcrum on which the scale of truth balances between evolution and creation. Since water is a key component to life and it is likely that water was involved in either evolution or creation, and since most creature lifeforms are dominantly made of water, it is logical and rational to assume that water played an important role in the initial creation events on day five. From here we are going to gloss over the very early part of those first amino acids and jump to the question of DNA formation—how did the first DNA form?

Was it Evolution or was it Creation?

Was the first DNA primitive, or was it more complete as we witness it today? This is where we really get to the core of that ultimate question, "was it evolution, or was it creation?" In this debate it really all comes down to the instructions that guide the reproduction of cells and the distribution thereof that ultimately form any particular creature. Since we know that creatures can morph to some extent, as witnessed in observations of our creature crossbreeding efforts in our modern era, we will not question if our modern idea of "speciation" could occur. This is obvious, but "species" are still a matter of the allowable scope of speciation of "kinds", and a matter of the specific definition used to designate a particular species.

While we do know from examination that DNA is alterable from simply observing the DNA of parents and their children, we don't know if there are deviation *limits*. DNA anomalies tend to correct themselves as is made evident when birth defects occur but are typically not passed on to the children in any visible way. While not absolute, this indicates a strong tendency of stability of strains of DNA or as the creation account puts it "kind", or Latin's "species", or Greek's "genos" all of which indicate some sort of group or family that share key attributes. We see little to defy this general idea, with the grasping exceptions illustrated in the various evolution charts. In referring to the evolution charts and their scope, the areas of those charts I refer to are the areas that align with divisions of the Biblical creation "kind" definitions where they have been artistically blended where little proof of actual creature blending exists. But again, I am speaking in strict adherence to the Biblical creation account which makes some very fundamental distinctions of form in regard to kinds where it indicates that kinds will produce like kinds. It neither supports nor denies any sort of long-term deviation ability of "kinds", thus scope deviation can neither be specifically argued for nor against from a Biblical standpoint.

For the sake of peace on this topic we will assume that enough deviation is possible for speciation within kinds to occur on the evolutionary model principle because we do know change can occur. But we do not know if there are limits, thus neither side can make a confirmable argument here and a great deal of modern debate comes down to scope and species definition. So to set that whole aspect aside, the question comes to DNA's initial formation. First, while we have what appears to be a pretty good idea of DNA's role in life, we do not yet know the fullness of DNA's purpose or function. DNA is a highly complex assembly of what appears to be what we would normally refer to as "data", but to refer to it as data implies intelligence in the assembly or design of DNA for better or for worse.

Let us imagine for a moment that DNA does evolve. We can then completely toss out the idea of creature evolution because it is the DNA instructions that evolve rather than the creature itself. This is because, to the best of our knowledge, the creature is cellularly assembled according to the instruction contained in the DNA strands. And because of this the creature will be what the DNA says it will be and few, if any, people on either side of this debate would argue against that particular point. We reach a point in the discussion of DNA where it becomes difficult to talk about without implying intelligence. This is because any sort of organization of information generally implies interaction of some type of intelligible guidance from an outside force. In the case of godless evolution, the outside forces are environmental, which is difficult to deny since we know that toxins can cause mutations in DNA, thus environment does provably affect DNA.

So now we will get into an even deeper aspect of this DNA dilemma. We can gloss over the entire system that surrounds DNA and focus on the DNA alone and make very compelling arguments that DNA could evolve into very different forms given enough time only due to godless environmental forces. But we are still faced with understanding the mechanisms that read the DNA instructions or sequence, where that reading then causes certain functions to occur that allow the cell that contains the DNA to make an identical copy of itself–including the DNA sequence itself. Here we are caught in a catch situation where the DNA instructs the design of the duplicate cells that will now read their own DNA strands to again make copies of themselves. This in itself is not the problem; the quandary we are left with is that since the DNA is the instruction for the DNA mechanisms' form, then how did that mechanism come to be in order to read the DNA that instructs the mechanisms' construction? Both the DNA and the mechanisms that read the DNA are used **together** to instruct cells to do their work. Is there a way to get past this dilemma? Evolution does so through incrementalization while at the same time glossing over the obvious engineering marvels of cells. Creation says the Creator did it. But we are still left with, how might this have occurred?

Getting back to the continents moving and weather patterns, the reason I mention that is because if temperatures need to be prime for life to thrive in either case of evolution or creation, then there is certainly enough variance in Earth's current climate that ranges from inhospitably cold to inhospitably hot and includes everything in between for life to begin in. And then add to that that the Earth has clearly gone through major geological changes over its history, changes that could have been better or worse for life to arise. The point to this is that since there are so many current variations, and if our continental movement assessments are remotely close to accurate, then the Earth has had a great variation of climate that would have had, and does now include, temperatures primed for life to explode–enter the egg!

What Came First, Archaeopteryx or the Egg?

I find it interesting that the creatures brought forth on day five favor the egg-laying type of birds and creeping creatures. What came first the chicken or the egg? That is to say the bird or the egg, and the lizard or the egg? The answer is both the egg and the creature.

At this point we are somewhat glossing over the DNA dilemma and skipping to the common assertion that the "kinds" mentioned in the creation account are what was initially brought forth by the waters. Assuming DNA is ready to go, could life form without a chicken or an egg? And what is an egg? An egg is protein, fat, and some minerals that are vital for life. This simple mix of readily available material inside the egg will produce most birds in less than a month when basic incubation conditions are met. The question here is, since plants came first and there are many plants that have fruit that produces fat and/or protein and essential vitamins, could something like that have been the source of material for the first bird? This is *not* implying that birds evolved from plants but rather much the way a tree is used as building material to build a house, organic plant material potentially was the source for the first egg-borne creatures to arise from without specifically needing a fully formed egg. This is offered only as a consideration and not as established fact, but it is far more plausible than the prominent evolutionary theory.

Plants are abundant and produce a wide variety of chemistry to support life, which is why we and animals eat plants. Plants at the water's edge are both in the water and contain water. Plants in these locations are some of the most nutrition-filled plants on Earth, in that many have a tremendous amount of vitamins and protein, fat, and carbohydrates. Such plants provide most everything needed for first life to form if the instructions for life are provided. Given the vast array of climate conditions that the Earth currently has and would have progressed through as the continents moved, we know that the proper incubation temperatures would be available somewhere on Earth at some point in times past. This can easily be proven by taking water temperature readings in various bodies of water in various climates, thus providing us with provable incubation temperature ranges needed for birds to grow and the provable nutrition and other basic ingredients for a bird embryo to feed off of. The shell would not be required; the incubating embryo would only need a still safe environment warm enough in which the creature could come to chick-level maturity. And since no other creatures existed they could easily do so without being disturbed.

So, now we have what are provable and available known temperature and nutritional requirements needed for birds and many types of creeping creatures and even fish to be brought forth. But we are left with the unanswered issue of instruction via DNA and fertilization. In this case, all needs are met which is supportive of both evolution and of creation, with the exception that it skips the plants' role in the evolution model which is explored in *The Science Of God Volume 2 – Day Three – Gravity, Land, Seas, and Evolution of Plants*. So all the way up to the point just before this in the creation account in the Bible, we can say with a great amount of certainty that the Bible is accurate, or at minimum cannot be legitimately scientifically discounted. Now when it comes to this bird-versus-egg issue, it is clear that the needs for egg-bearing creatures to arise can easily be met without an actual egg from a bird, but we lack fertilization and the very important DNA from that fertilization. Here we will pause the day five creatures for the moment and move on to day six.

On day six is it the *Earth* that brings forth. Since land "beasts" are typically mammals and are born from a female of that "kind", we have to wonder, what came first, the adult, or the baby? Much like the chicken-versus-egg issue, we have to wonder if a baby mammal creature could somehow come first. With mammals it is more difficult to see a clear path without a parent because mammals are attached to the mother with the umbilical cord and through that umbilical cord they are given the nutrition and other materials needed to grow. The same general needs for a baby mammal can be met in much the way the bird embryo needs are met. But how could those nutrients get to an embryo that is normally connected to the mammal without it being connected to a mother via an umbilical cord?

Eggsactly

To answer this we can look at photos of chick development in an egg. There are great photographic examples for the day-by-day chick incubation progress where the egg has been opened so you can clearly see the progress of the chick's development. Search a library or online for "chicken egg development" and look for progress pictures. Doing so is very informative! You will notice that on the yellow yolk liquid there will be a white spot. This whitish spot is the point of fertilization that occurred while the egg was forming in the hen. The white spot point of fertilization takes about a one day or about twenty-four hours to form. Once the egg is laid, it needs to be incubated by the hen sitting on the egg. However, with artificially incubated eggs it is done at a temperature of just below one hundred degrees Fahrenheit. Incubation periods vary in the various breeds of birds, but chicken incubation is only about twenty-two days. Many small birds only take about twelve days to incubate and hatch. This is all very important to see and understand, and if possible, to watch.

Here is where things get really interesting: Birds do have belly buttons, or maybe better stated navels. Birds are attached to the yolk sac in much the way a baby mammal is attached to its mother. While birds do not have a long umbilical cord, they do draw all of their nutrition from that central location of connection. This point of connection appears to be the point of fertilization, and within the first day blood is being created. The heart actually begins to beat in as little as about thirty hours, and blood vessels and veins are forming at the same time. Much of this is clearly seen in photographic evidence of the progressive stages of chicken egg incubation. Look for these pictures, they are easy to find in our modern era and are quite revealing.

What does all of this mean, and why is it important in the evolution-versus-creation debate? Since egg-borne creatures and mammals both have a central location for the infusion of the needed elements and nutrients for life for begin, grow, and thrive, and since the life of a bird such as a robin or sparrow can begin and hatch inside of two weeks and in another two weeks—fly, the only thing we would need for shell-less eggs to quickly bring forth life is for the provable existent conditions and materials to be present. All of that is verifiably, falsifiably, and provably available. At that point given that DNA was presented, a creature could form in as little as two weeks and this could be done in abundance wherever the required conditions were met, which would be a wide variety of locations *all around the globe.*

This same model can be applied to mammals, especially those with very short gestation periods, such as gophers that have a three-week gestation period. All of the same principles apply to mammals that apply to egg-borne creatures. You might wonder about the issue of Earth bringing forth versus the waters bringing forth. In this case it would likely be the raw earth materials available to the embryo in the day six creatures brought forth by the Earth. Day six creatures would also likely have been brought forth in wet or damp conditions but with very different source materials to meet the needs of those embryos. In this theory of egg-like conditions for creature life to get started, it is important to note that it would be vast and robust, and it could occur anywhere the needed conditions were met, which would be a wide-ranging variety of raw nutrients to be utilized as egg-like materials all around the globe.

Shared Attributes

The fact that mammals and egg-borne creatures share a likeness in their early stages is explained by the fact that they both use a centralized development mechanism as explained in previous chapters. The earlier in the gestation process then the lower the cellular resolution is, thus the more similar the creature kinds and their organs will appear in the early stages of development. This particular use of common-function is a very strong indicator of intended function, meaning that it is the result of engineering of some sort. I am not asking anyone to take this as fact, but rather to consider its ramifications. The ramifications are that life could spring up *nearly everywhere,* and since the conditions would vary, it would vary the output of the material, that is to say the creatures arising out of the waters and arising from the Earth. All of this is consistent with what we see in nature, and with what we see in the Biblical creation account, in geology, and in science.

At this point we are stuck with one persistent unanswered area that keeps sneaking up on us regardless of being evolved or created: How did any DNA sequence come about? This is essentially asking how could any naturally occurring "egg" raw materials become fertilized? Let's ask:

Dear loving God and Creator,
How did the first creatures form? How did their DNA come about? Did they get fertilized, and if so, then how? Did the entire DNA string of any one creature type or kind come from fertilization? Or was it some combined source? If solely from fertilization, then where did the point of fertilization come from? If from shared sources then what are those sources and where are they from? These are our questions.
Dear Lord God Heavenly Father we worship You and praise You and glorify You and we bless You who is our wondrous Creator. Let us know You better mighty Creator through Your science. We worship and honor You and give You thanks and praise and we will be greatly blessed if You deliver clear and concise answers to these questions for all of us humans. Send me Lord, we will share Your truths with the world. We bless You oh my Lord God and Creator!
Your Faithful Servants

I have to relay this to you. Immediately upon completing typing that brief prayer asking for answers and thanking God, I was urged to come and look out the window at a very amazing rainbow. It was one of the most vivid and complete rainbows I have ever seen. It was a double rainbow and it was so full that near the bottom it actually curved inward a bit. It was really very beautiful. Here is an interesting thing about rainbows that few people know. Rainbows are said to be a promise that God would never again destroy the Earth and the life on it with a flood. Most people know that, but what they don't know is that the letter "B" also makes a "V" sound in Hebrew and in several other languages, thus "rainbow" can be pronounced rainvow. That's a rain *vow* as in—I promise you.

Life in general is robust, and from our human experience it is unstoppable! Life happens quickly, it is adaptable, and it is versatile. Life is consistent and persistent, and it wants to live. Life will defend itself voraciously whenever it is able and needs to do so. There is what seems like a limitless variety of proper raw materials and locations to give rise to a wide variety of creatures simultaneously, thus bringing into question any need for single-source life morphing in to more complex life as unnecessarily demanded in evolutionism. Everything proposed here is readily available and possible, but we are still stuck with the issue of fertilization.

What is fertilization? Since DNA is built from a complex series of four simple parts or bases, cytosine "C", guanine as "G", adenine "A" and thymine as "T", it is those four bases that form the DNA sequence, but what guides the various sequences that offer each creature kind its unique form? And, is this instructed from the fertilization source or from the egg source? In humans we are quite certain that the DNA sequence is a combination of the egg-source *and* the fertilization-source. We are also quite certain that the same is true of most creatures. Since DNA is most likely a combination of egg-source and fertilization-source materials, life would have great diversity from the egg-like natural raw source material alone basically borrowing some of its DNA from plants. The instruction set from fertilization would not have to hold all of the data, but rather only some generalized creature-form data.

The Spark of Life

If you take the time to look into the moment of conception you will find that there is a physically visible spark that occurs at specific moment that the single very tiny sperm cell breaches the egg's outer layer. What exactly is occurring at that specific moment in time? It is an exchange of data. What we don't yet know is if the spark is some from a physical stress reaction, or if it is a means by which the data from the fertilization source is added to the egg source.

Going out on a tree limb here, we see certain random patterns in nature, trees and blood vessels and large lightening arrays all have a similar branching. This unique structuring shows resistance patterning. It is a type of randomized patterning. We know without question that data can be passed through "waves" of energy and through pure electrical energy. We do this constantly with all of our various electronic activities for radio waves, and in doing so, we pass vast amounts of data. So the question is, since the Creator creates from mind alone, how could any intended creature design get from the mind of God to the raw materials?

The book *Understanding Prayer – Why Our Prayers Don't Work – The Prayer How-To Manual*, explains some basics of certain medical equipment and other experimental equipment that can both detect thinking or brain activity, and can also affect or inhibit the brain and therefore inhibit our movements. The book goes on to discuss how humans can communicate in a subconscious manner through what is basically electricity. That duplicatable laboratory communication evidence just mentioned indicates that it is not only possible, but provable, that data can be transferred through electricity. Lightning is electricity and can carry with it data. Since our brains can provably interact with and be interacted with via electrical energy, we can safely assume that data could be exchanged through electricity and thus through lightning.

We know that the moment of conception creates a spark via repeatable observation, and we know that data can be transferred through electricity. We also know that electricity can affect the brain and that the brain can affect electricity without anything actually physically connected to the brain, thus using basic human logic, we know that our process of thought or thinking affects electricity. All of these things are verifiable, falsifiable, provable, testable, observable, and repeatable. Electricity is a means of passing data, so we likely pass data of some sort when we are in a state of joy or a state of anger, which explains why we sometimes get negative "vibes" from someone or we feel happy in the presence of someone.

If a Creator exists, then that creator has to do everything through thought. Since there is no visible giant man in the sky to build all things, and since according to Genesis, the Creator created all things, it is logically apparent that this Creator has no body and is logically *Powerful Eternal Thought*. This logically demands that the Creator created by thinking it, much the way we do when we imagine something that never was. We developed and can see it in our mind and imagine all of the details about it. We can do this to such a point of precision that we are then able to write those instructions down to be repeatedly followed at some point in time by others. When those instructions are then implemented, our thoughts are made into tangible substance using other substance to create the final outcome. While someone could argue this and say that we create with our hands, the truth is that all creating first occurs as a thought or concept in our mind. Upon forming that concept it now actually exists in thought and only you can erase it, but that is actually impossible, because once it is thought by you, you may try to suppress it, but that thought can re-emerge at random times in the future, proving that it still exists.

Then to take this mental creation to a higher level, we write instructions in a physical way, but what we need to understand is that those instructions are passed electrically through our bodies from our thought/mind to our brain then to our nerves and muscles to allow us to perform the physical actions required to record the instructions or to actually follow the instructions. This is also verifiable, falsifiable, provable, testable, observable, and repeatable.

All of the points I am making have been scientifically proven with the exception of the egg raw material point that was made in this chapter. This brings us again to this point of DNA. What is the source of programming for sequencing the known four simple DNA bases of cytosine "C", guanine as "G", adenine "A" or thymine as "T"? Since the Bible's Genesis One creation account cannot be disproven because it is all scientifically accurate as described in this book and is verifiable and falsifiable through basic human logic and scientific analysis and testing, and it also does not specifically defy evolution, it must be taken seriously by any rational mind. I am not saying someone must believe it all; I am saying that to discount it on a scientific basis is utterly irrational.

Regardless of evolution or creation, what is the source of DNA sequencing? What could have given rise to DNA? In the evolution theory it is lightning striking chemical gases that create amino acids. Could this be true? Yes it could be true, because it has been done in labs with scientific experiments, but it is not DNA, it is only amino acids at that point, and it has no observable DNA sequencing. When the needed source materials are available and excited with electricity, synthetic amino acids are created. The implication is that since we intelligently can do this in a lab using the thoughts in our heads and deduce such an experiment through the electrical impulses from our thoughts to our brain which then goes from our brain to our hands, lightning could also do so in a random manner if it were to, by chance, happen to strike the proper chemicals needed to create the amino acids. Is it rational to imagine that this could happen by accident and then from that, over billions of years creatures eventually evolved?

Actually, no, it is not rational to believe that. Evolution suffers from one key problem: Evolution has to invent a means for *every* causal action that would have affected the structure of the lifeform that was developing in that theory all along the way up until our current experience of creatures. This is where evolution fails, and at each point of failure, and there are many, it is forced to gloss over that which it cannot explain. To compensate for the inability to explain each critical step of the early stages of the formation of life, evolutionism simply adds time, and lots of it!

There are mechanical design issues that an honest person examining life cannot ignore. There are too many intricate mechanisms in any cellular structure to claim that it happened by random chance. With each unexplainable cell-machine component, the concept of *random* adds an exponentially large factor to the odds of it all happening through random or chance occurrence. The childish monkey typing an infinite amount of characters, spoken of earlier in this book doesn't hold up to scrutiny because that obviously will never happen, and even if it could in theory happen, I then would have the originator of that ridiculous theory have their monkey monkey around with the raw ingredients to life to see if their theory-monkey could stumble upon life in the same way as the *Hamlet*-theory-monkey stumbled upon a perfect copy of *Hamlet* in and infinite amount of time. Cellular design is incomparably more complex than is *Hamlet*. Any person who is being honest and using true logic is immediately going to understand that the odds of that monkey doing either case is *zero*. Further, cells cannot duplicate without all of the cell components existing, so evolutionism is forced to explain how those mechanism were produced in the very first duplicatable cell.

If a Creator exists then that Creator would be nothing except pure mind and pure thought able to "create" and initiate creature life using electricity generated through a static charge of the elements, and could infuse that static charge of lightning with instruction using *mind* and *thought*. It is undeniable that when lightning strikes the raw material "waters" and "earth" that data could easily be passed. Lightning is the spark of life and in it are the instructions that are inscribed into the raw materials. The lightning causes the amino acids and delivers the instructions for the organization of the C, G, A, and T bases that form DNA– without them DNA does not exist. This fundamental "kind" instruction-data causing, and delivered to, the amino acids, together with the plant DNA makeup, allows for waters and earth to simultaneously bring forth, all over the globe, a wide variety of the general "kind" forms specified in the Genesis One creation account, which could potentially even include variations of those "kind" forms.

Instead of blindly accepting the illogical rationale that random occurrence caused everything; let us at least consider that the same lightning that was hijacked by evolutionism, is capable of delivering a sequence of deliberately intended instruction for the C, G, A, and T bases to become ordered and used in combination with the raw material DNA from the waters and the earth to quickly form life *without any* evolution required.

This is not to demand that anyone must accept this view, but rather, the only way that the creation account can be found unacceptable is if someone has specifically and deliberately chosen to illogically reject the Creator in the face of the overwhelming evidence that illustrates otherwise. Lightning delivering instruction to raw materials causing organization of C, G, A, and T DNA code that is able to form life, allows for many of each "kind" to be brought forth quickly all at the same general time. This allows for the creatures' two genders to be brought forth in parallel time or in a near approximate time frame.

Life occurring in this way would allow the "waters" and "earth" to quickly, in typical gestation periods, bring forth a wide variety of fully formed creatures for each "kind" mentioned in Genesis

One. DNA shows *guided* complexity and this explains it well. This also allows for a limited amount of evolution over time, but it is bound to the limits of each "kind" instruction established in the DNA. This also allows for days five and six to be either long or short. The Genesis account says nothing about the age of the creatures or the maturity of the creatures, it just says that they will be of a "kind" and will be flying or creeping. Genesis does *not* say what most people have been taught it says. When read with an open mind in an authoritative version, Genesis One is far more accurate and scientific than most people think or are willing to accept or are allowed to hear.

The advent of the creatures was not *evolution* it was a *revolution* of deliberate interactive creativity and the prompt arising of diverse *kind* life.

Creature Life is a Revolution!

Announcements

Theoretical Physics for Everyone!

Can we go back in time? Can we break the light-speed barrier? Is our Sun turning into a black hole?

Explore the secrets of the Universe with a new approach to science that sheds a greater light on pop-science. Mysteries of the Universe are revealed in this easy to understand insightful, in-depth, and thought provoking book about the science of astrophysics.

Theoretical physics can be more than mere theory when the theory is sound. You don't need to be a rocket scientist to understand most of physics; everyone is welcomed in the quest to discover the mysteries of the Universe!

This breakthrough book exposes errors of modern science in the same way that Copernicus, Galileo, and Newton did centuries ago. Are Einstein and Hubble amongst the group of gifted minds that set forth our understanding of the Universe, like those of centuries past. Or are Einstein's and Hubble's theories wrong? Explore these and other questions in *Bending The Ruler - Time Travel, The Speed of Light, and Gravity* and learn how to become one of the great minds that discover the mysteries of the Universe!

Search: Bending The Ruler Book
SayItBooks.com

Announcements

Take Back Control of Your Life.

If you feel stuck while life unfairly drags you down, then now is the time to take command of your life and learn how to overcome the source of your troubles.

Those around us are often those who hold us back from living rich and robust lives. Realizing that those around us are often those who hold us back helps us to understand somewhat, but in order to free ourselves from their grasp and break the chains that bind us, we need to know *why* this happens.

Cut to the core of your problems with *Hot Water* as it walks hand-in-hand beside you through each detail of the cause of problems while exposing the dirt that society buries us with. This thought provoking book explains the details and how most troubles come to be so that you can better understand what to do about it, allowing you to take the control of your life away from those around you to place it firmly back into your own hands where it belongs.

Advance to your next place in life and richly and robustly live your life filled with wealth and joy. *Hot Water And Your Perceived Identity* assists in gaining full control of your life to change your future forever!

Search: Hot Water Book
SayItBooks.com

Announcements

MARRIAGE MANUAL
MAKE YOURS A
Red Hot Marriage
Made In Heaven Filled With Passion and Joy

Learn the Secrets to a Successful Marriage

Have you been trying unsuccessfully for years to tell your spouse the way you truly feel? Are you suffering in a lackluster marriage? Is your marriage on the rocks? Are you planning on getting married in the future? If you answered yes to any of these questions then *Red Hot Marriage* is for you! This straight-forward book covers these and many other common marriage problems and also reveals the causes and solutions for some problems that are not-so-common.

The information in this powerful book, like a true friend, can be at your side with each step you take in restoring your life and relationship to where you likely imagined them to be.

We all deserve lives filled with joy and passion, but our relationships have been tainted by society and by our upbringing. *Red Hot Marriage* strips away all of the lies that we have been inadvertently taught, and quickly teaches you how to regain control of your marriage so that it can be as robust, fulfilling, and passionate as you expected. The mysteries unveiled in *Red Hot Marriage* can have you in command of your marriage in short order as friends and family watch in amazement while you and your spouse walk the path to a strong, vibrant, healthy *Red Hot Marriage*!

Search: Red Hot Marriage Book
SayItBooks.com

Announcements

THE FAMILY MANUAL
HOW TO BUILD A
STRONG FAMILY
A FOUNDATION OF ROCK

Building A Strong Family Is Easy When You Know!

The world has us believing that, somehow, we are now different than people were as little as fifty years ago. With all of the emphasis on modern behavioral disorders and the mass misdiagnosis of pop-culture diseases, parents have few places to go for information that is true, insightful, and trustworthy.

Strong Family explains, in detail, how family life slowly becomes tainted to a point where our children too often become rebellious and, sometimes, even unmanageable. This even happens to parents who are very loving people.

How to prevent these issues from occurring in the first place is explained in *Strong Family*. But more importantly, *Strong Family* explains the details about how to stop it from progressing further and even how to reverse the damage. *Strong Family* takes a no-nonsense approach to revealing the secrets and mysteries of how parents raise smart, productive, healthy children.

We all deserve joy and love in our family life. Intelligent, healthy, kind children are a right that all parents have, but without understanding the details explained in *Strong Family*, the quality of your children is left to chance and your rights are forfeited. Don't roll the dice with your family. If you want to know the secrets to unlock the mysteries and solutions to a great and joyful family, then *Strong Family* is for you!

Search: Strong Family Book
SayItBooks.com

Announcements

The Prayer How-To Manual
Understanding Prayer
Why Our Prayers Don't Work

Learn the Real Secret of Prayer

There's a secret that many have tried to understand but failed to accomplish. We pray day after day after day with little or no positive results, causing us to lose faith.

Some people believe that there's a secret method that must be followed to get your prayers answered and receive the things you want in life, but their success is limited, if it comes at all; while others believe that they're not worthy to have their prayers answered. Few people know the True secret, and when they tell us we often misunderstand them.

Understanding Prayer explains, in easy to grasp language, the mysteries behind many causes of prayer failure. True success in your prayers is not measured by how often you pray, how long you pray, or even how badly you want something and how hard you for pray it. True success in your prayer life is measured by *results*!

Understanding Prayer offers you the opportunity to get those results as it reveals the mysteries of a full and robust prayerful connection allowing you solid and repeatable results nearly on command. A little time to read and pray is all it takes to quickly put these sound, true, simple principles to work for you and your family. Gain the understanding of prayer and of how to receive the blessings of financial and mental wealth that can benefit you and keep you free from strife and trouble for years to come!

Search: Understanding Prayer Book
SayItBooks.com

Announcements

When You Dream... DREAM THIN™
The Weightloss Repair Manual

Learn How to Lose Weight While Sleeping

How many people do you know who exercise and still can't seem to lose weight? Has that ever happened to you? As a matter of fact, because we don't know the vital secrets that are shared in *Dream Thin*, many of us actually end up *gaining* weight when we exercise.

Do you hit your weight loss goals? And does your weight stay off when you do actually lose some weight? Even many doctors miss the *real* answers to weight loss. If you doubt this, then simply look at the waistlines of many medical doctors and nurses.

Weight loss is easily mastered when you understand a few basic principles. We often go on fad diets or follow the orders of our doctors, only to put the weight back on even faster than we lost it. Many of us suffer from unnecessary disease, and some of us will die too young.

Dream Thin does more than simply share answers to weight loss mysteries. *Dream Thin* explains the important details of *why* and *how* weight loss connects to mind and body. The information in *Dream Thin* allows you to make weight loss permanent without having to try so hard. Don't make more of the same empty promises to yourself each New Year's Day. Instead, quickly and easily change things today and make all of your tomorrows better with *Dream Thin* while still enjoying all of the foods you eat today—and yes, even fast foods!

Only you can choose if you want spend your hard-earned money on medical bills and funerals, or if you would rather spend your time and money looking great while being out and about and enjoying life with friends and family as intended!

Search: Dream Thin Book
SayItBooks.com

Announcements

How to Win When You Think You've Been Beat

Feeling grateful is a bit of a struggle when we face tough times in our own lives, and avoiding depression during those times can be tricky. The world cares little of us when we face our own personal struggles, in fact life kicks us when we're down. You've probably experienced the world caring little of your past or present problems, so looking to "The World" for rest and peace is typically of little help.

It doesn't have to be this way! You can change your disposition, and thus, change your future! It's no big secret and it's not difficult, but "The World" won't tell you that, so very few people ever get to hear or understand this simple "secret".

It's amazing to see the people and situations you can attract into your life when you find your own proper perspective, and once you find it you will not want to let it go! Days that test you to your limit become far easier to overcome, making every future test easier than it otherwise would have been.

Simply understanding a few key basics can change your direction in life in short order and can make life a whole lot more peaceful and Joyful! Let *Thank You God – Finding Gratitude in Hard Times* be one of your keys to peace and Joy!

Search: Thank You God Book
SayItBooks.com

Announcements

Volume 1 - The First Four Days

Is there a God? Did we evolve? Did everything start from a big bang? These questions have been plaguing our minds for many years. Only science-minded people and clergy seem to have the answers. But do they really have any true answers?

Is what we are told by science true? Is what we are told by the Church true? Or are there other better explanations for everything? Did we hitch a ride from Mars, or is that all fantasy science? Was everything created in six twenty-four hour days, or did it all take billions of years to happen? Few people are willing to even fully consider these questions, and even fewer have any coherent answers. *The Science of God Volume 1 – The First Four Days* challenges your current beliefs while asking tough questions of science and of the Church.

For years, Christian after Christian has attempted to argue for God and the Bible's Creation only to fail miserably. Why is this, why is it that Christians cannot seem to win this debate? Often Christians think they are winning the debate only to find themselves at a loss to answer the real questions, and then they get mocked for their poor answers.

Whether you are a scientist or an average Christian and want to discuss the Creation debate, *The Science of God Volume 1 – The First Four Days* is a mandatory read for you. *The Science of God* takes you through the thought process to enable you to speak intelligibly about Creation, the cosmos, evolution, and astrophysics.

Search: The Science Of God Book
SayItBooks.com

Announcements

Discover the Building Blocks of Life

Are you feeling confused about how plants really came to be? Did they evolve, or did some "God" create them? Could they exist without the sunshine? When did gravity begin, and does that matter regarding the arrival of plants? If a God did create the plants, then exactly how might that have occurred? Or if the plants evolved, then what did they evolve from?

There are many questions that need to be answered, but who has the time to study these things in depth? After all, scientists do this as a fulltime career and even they lack many answers to such questions.

The Science of God Volume 2 - Gravity, Land, Seas, and the Evolution of Plants offers unique perspectives to assist in quickly discerning the onslaught of information from both the religious and scientific sides of this debate. While there are some scientists and religious people who attempt to stand on both sides of the evolution versus creation discussion, doing so often harms their credibility due to conflicts in their logic.

The Science of God Volume 2 - Gravity, Land, Seas, and the Evolution of Plants stands alone in explaining and answering the central questions that many people have surrounding the topic of plant evolution versus creation.

**Search: The Science Of God Book Volume 2
SayItBooks.com**

Announcements

Rocking the Cradle of Life
A Decent Account of Descent

Have you ever wondered if humans actually did evolve from apes? Or maybe, if we were specifically created, then how might have that occurred? There sure are a lot of opinions on the evolution versus creation topic. And too often these views use confusing technical jargon that few people care to learn or have ever even heard.

The answers to the questions you might have are, in many cases, the same answers that many other people seek. When you have solid answers that are difficult for someone to thwart, it's good to share those answers so that others can also feel confident with their own understanding of the arrival of mankind and the level of importance that it has in their own lives.

The Science Of God Volume 4 - Evolution versus Man – In Our Image takes a deep but simple dive into the human evolution versus human creation debate using simple language that everyone can understand and enjoy!

If you have thoughts that you have been reluctant to share, then suspend your thoughts for a bit and open your mind to consider the perspectives and evidence presented in *The Science Of God Volume 4 - Evolution versus Man – In Our Image*. You will acquire a much clearer view of the subject as you read the various points made in this engaging book about the arrival of mankind.

Search: The Science Of God Book Volume 4
SayItBooks.com

Announcements

Rock the Boat with Layers of Truth

Do you believe that the entire world flooded roughly four thousand years ago and that a man named Noah built a large boat to save a small remnant of human and animal life that would repopulate the entire Earth? This is the belief of many Christians, Jews, and Muslims, but then we have those who believe that the entire story was written thousands of years ago for entertainment only.

Could either case be true? Is either realistic? After all there is a lot of evidence of catastrophic worldwide flooding. But then there are those making the point that there's not enough water on Earth to cover the mountains. So, which, if either, is it? If either case were proven to be undeniably true it would have major impact on opposing perspectives. If it never occurred, it would devastate most Bible-based religions. But how would it affect modern sciences if it was proven true? It would force every scientist to face a reality for which they have not been educated.

Take a journey through these and other Biblical flood questions and consider the perspectives presented in *The Science Of God Volume 5 – Boats, Floods, and Noah – The Deluge*, a truly logical scientific explanation of the viability regarding the Biblical flood of Noah's time.

Search: The Science Of God Book Volume 5
SayItBooks.com

Announcements

The Cornerstone of Moral Civilization

Was Jesus really the "Savior"? Did Noah really save humanity from extinction? Did Adam and Eve really get evicted from the Garden of Eden? And what does the word "Bible" mean anyway? When studying or even just reading the Bible, many questions arise to a point where the Bible can be confusing. But when you have certain information before you begin reading, it can instantly propel you to a deeper level of understanding by nothing more than knowing a few key points.

It takes people years to realize some of this information, yet it's not some big secret that only scholars and theologians know. No, this information is for everyone and it's easy to grasp these pieces of information about the Bible and some of the events described within it. Be prepared to have your current views challenged because many things are not as we have been taught.

To truly Understand the Bible, we must open our minds and toss aside all of our biases. Knowing and grasping the often-unrealized basic information presented in *Understanding The Bible - The Bible How-To Manual and The Things We Don't See* brings the Bible to life in a way that shows you, personally, its undeniable relevance to the world, to our culture, and to your very own life!

Search: Understanding The Bible Book
SayItBooks.com

Announcements

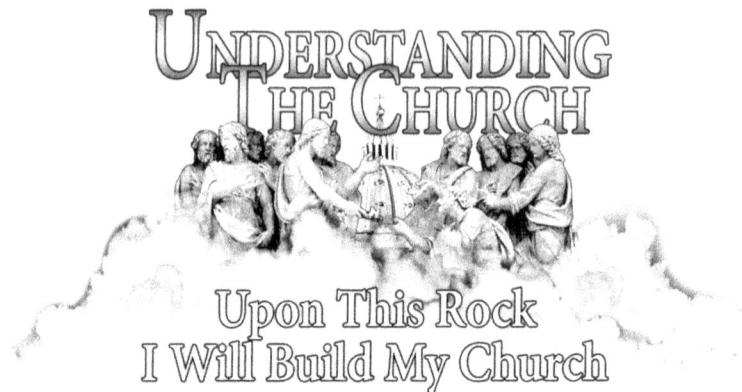

Church in the Lurch - a House Built Upon Sand

The Church is rapidly dying, and much of the clergy in recent times have been doing it more harm than good. People are fleeing from the Churches as they seek a religious perspective that fits a modern worldview. Should we revive this old Church and try to save it from its own demise? What exactly is "The Church", and who or which of the many religions is the official caretaker of it?

The Christian religions of the world have done their fair share of damage to themselves and to the world, but in the bigger picture, they have done more good than damage. Saving the Church is probably worth our collective efforts because the Churches are perhaps the most charitable group of organizations that existed throughout history and even up to today.

The main reason that the Churches are in the rough condition that they are today is due to a lack of understanding by clergy and congregation. We can overcome this dark era of the Church and revive it only through *Understanding The Church*.

Understanding The Church will help you in Bible study, or even to simply better understand the Church. But most importantly, *Understanding The Church – Upon This Rock I Will Build My Church* will help to revive this dying patient.

Search: Understanding The Church Book
SayItBooks.com

Notes

Notes

Notes

Notes

www.ingramcontent.com/pod-product-compliance
Lightning Source LLC
Chambersburg PA
CBHW071658090426
42738CB00009B/1579